Coastal Disasters

Series on Coastal and Ocean Engineering Practice

Series Editor: Young C Kim
(California State University, Los Angeles, USA)

Series on Coastal and Ocean Engineering Practice – Vol. 3

Coastal Disasters

Editor

Young C Kim

California State University, Los Angeles, USA

World Scientific

NEW JERSEY • LONDON • SINGAPORE • BEIJING • SHANGHAI • HONG KONG • TAIPEI • CHENNAI • TOKYO

Published by

World Scientific Publishing Co. Pte. Ltd.

5 Toh Tuck Link, Singapore 596224

USA office: 27 Warren Street, Suite 401-402, Hackensack, NJ 07601

UK office: 57 Shelton Street, Covent Garden, London WC2H 9HE

Library of Congress Cataloging-in-Publication Data

Names: Kim, Young C., 1936–   editor

Title: Coastal disasters / editor, Young C Kim, California State University, Los Angeles, USA.

Description: Singapore ; Hackensack, NJ : World Scientific Publishing Co. Pte. Ltd., [2026] |
    Series: Series on coastal and ocean engineering practice ; vol. 3 |
    Includes bibliographical references.

Identifiers: LCCN 2025036266 | ISBN 9789819816576 hardcover |
    ISBN 9789819816583 ebook for institutions | ISBN 9789819816590 ebook for individuals

Subjects: LCSH: Coastal engineering | Coastal zone management | Shore protection |
    Emergency management

Classification: LCC TC209 .C63 2026

LC record available at https://lccn.loc.gov/2025036266

British Library Cataloguing-in-Publication Data

A catalogue record for this book is available from the British Library.

For any available supplementary material, please visit
https://www.worldscientific.com/worldscibooks/10.1142/14404#t=suppl

Desk Editors: Soundararajan Raghuraman/Steven Patt

Typeset by Stallion Press
Email: enquiries@stallionpress.com

Preface

Coastal and Ocean Engineering Practice Series, Volume 1, was published in 2010, and *Design of Coastal Structures and Sea Defenses, Volume 2*, was published in 2015. After that, the project became dormant. Recently, Mr. Steven Patt asked me to reactivate the series. Hence, *Coastal Disasters, Volume 3*, was born in 2024.

Coastal disasters are physical phenomena that expose a coastal area to the risk of property damage, loss of life, and environmental degradation. This book, *Coastal Disasters, Volume 3*, provides the most up-to-date technical advances on disasters caused by hurricanes, coastal floodings, tsunamis, sea-level rise, and beach and bluff erosions.

Written by renowned practicing coastal engineers, this edited volume focuses on the cause and effect of coastal storms, seismic events, impacts of climate change, and shoreline change. This book is an essential source of reference for professionals and researchers in the areas of coastal, ocean, civil, and marine engineering.

I would like to express my indebtedness to six authors and co-authors who have contributed to this series. It has been a year-long project, and their support and sacrifices have been deeply appreciated.

Finally, I wish to express my deep appreciation to Mr. Steven Patt of World Scientific Publishing who gave me invaluable support and encouragement from the inception of this series to its realization.

Young C. Kim
Los Angeles, California
April 2025

About the Editor

Young C. Kim, Ph.D., Dist.D.CE., F.ASCE, Honor.m.IAHR is currently a Professor of Civil Engineering, Emeritus at California State University, Los Angeles. Other academic positions held include a Visiting Scholar of Coastal Engineering at the University of California, Berkeley (1971); a NATO Senior Fellow in Science at the Delft University of Technology (1975); a Visiting Scientist at the Osaka City University (1976); and a Visiting Professor at Universite de Nice-Sophia Antipolis (2011–2013). For more than a decade, he served as the Chair of the Department of Civil Engineering (1993–2005) and was the Associate Dean of Engineering in 1978. For his dedicated teaching and outstanding professional activities, he was awarded the university-wide Outstanding Professor Award in 1994.

Dr. Kim was a consultant to the U.S. Naval Civil Engineering Laboratory in Port Hueneme and became a resident consultant to the Science Engineering Associates where he investigated wave forces on the Howard Doris platform structure, now being placed in the Ninian Field, North Sea.

Dr. Kim is the past Chair of the Executive Committee of the *Journal of Waterway, Port, Coastal and Ocean Division* and served on the Executive Committee of the Technical Council on Research and the Civil Engineering Department Heads Council of the American Society of Civil Engineers (ASCE). He also served as the Chair of the Nominating Committee of the International Association for Hydro-Environment Engineering and Research (IAHR). Since 1998, he served on the International Board of Directors of the Pacific Congress on Marine Science and Technology (PACON). He is the past President of PACON.

He has been involved in organizing 14 national and international conferences. He has authored 8 books and published 56 technical papers in various engineering journals. Recently, he served as an Editor for the *Handbook of Coastal and Ocean Engineering* (2010), *Coastal and Ocean Engineering Practice* (2012), *Design of Coastal Structures and Sea Defenses* (2015), and *Handbook of Coastal and Ocean Engineering, Expanded Edition* (2018), which were published by the World Scientific Publishing Company. He is a Distinguished Diplomate of Coastal Engineering from the Academy of Coastal, Ocean, Port and Navigation Engineers (2011), a Fellow of the American Society of Civil Engineers (2012), and an Honorary Member of the International Association for Hydro-Environment Engineering and Research (2019).

List of Contributors

Mary A. Cialone
Coastal and Hydraulics Laboratory
U.S. Army Engineer Research and Development Center
Vicksburg, Mississippi
mary.a.cialone@usace.army.mil

Steven W.H. Hoagland
Department of Civil and Environmental Engineering
Virginia Tech
Blacksburg, Virginia
University of Tennessee
Knoxville, Tennessee
shoagland@vt.edu

Yukiyoshi Hoshigami
Professor
Department of Oceanic Architecture and Engineering
Nihon University
Chiba, Japan
hoshigami.yukiyoshi@nihon-u.ac.jp

Jennifer L. Irish
Department of Civil and Environmental Engineering
Virginia Tech
Blacksburg, Virginia
jirish@vt.edu

Catherine R. Jeffries
Department of Geosciences
Virginia Tech
Blacksburg, Virginia
catherinej@vt.edu

Catherine M. Johnson
National Park Service
University of Rhode Island
Narragansett, Rhode Island
catherine_johnson@nps.gov

Timothy W. Kana
Principal Emeritus
Coastal Science & Engineering, Inc
Columbia, South Carolina
tkana@coastalscience.com

Nobuhisa Kobayashi
Professor and Director
Center for Applied Coastal Research
University of Delaware
Newark, Delaware
nk@udel.edu

Jirat Laksanalamai
Center for Applied Coastal Research
University of Delaware
Newark, Delaware
jiratlak@udel.edu

Kyle Mandli
Department of Applied Physics and Applied Mathematics
Columbia University
New York, New York
kyle.mandli@columbia.edu

S.A. Sannasiraj
Professor
Department of Ocean Engineering
Indian Institute of Technology, Madras
Chennai, India
sasraj@iitm.ac.in

Vallam Sundar
Professor Emeritus
Department of Ocean Engineering
Indian Institute of Technology, Madras
Chennai, India
vallamsundar@gmail.com

Sean Vitousek
United States Geological Survey
Pacific Coastal and Marine Science Center
Santa Cruz, California
svitousek@usgs.gov

Robert Weiss
Department of Geosciences
Virginia Tech
Blacksburg, Virginia
weiszr@vt.edu

Yalcin Yuksel
Professor
Department of Civil Engineering
Yildiz Technical University
Istanbul, Turkey
yalcinyksl@gmail.com

Tingting Zhu
Center for Applied Coastal Research
University of Delaware
Newark, Delaware
ztting@udel.edu

Contents

https://doi.org/10.1142/9789819816583_0001

Chapter 1

Advances in Morphodynamic Modeling of Coastal Barriers: A Review

Steven W.H. Hoagland[*,**], Catherine R. Jeffries[†,††],
Jennifer L. Irish[*,‡‡], Robert Weiss[†,§§], Kyle Mandli[‡,¶¶],
Sean Vitousek[§,‖‖], Catherine M. Johnson[¶,***]
and Mary A. Cialone[‖,†††]

[*]Department of Civil and Environmental Engineering,
Virginia Tech, 750 Drillfield Dr, Blacksburg, VA 24061, USA

[†]Department of Geosciences, Virginia Tech,
926 West Campus Dr, Blacksburg, VA 2406, USA

[‡]Department of Applied Physics and Applied Mathematics,
Columbia University, 500 W 120th St, New York, NY 10027, USA

[§]United States Geological Survey, Pacific Coastal and Marine Science
Center, 2885 Mission St, Santa Cruz, CA 95060; Department of
Civil and Materials Engineering, University of Illinois at Chicago,
842 W. Taylor St, Chicago, IL 60607, USA

[¶]National Park Service, Region 1; University of Rhode Island,
Bay Campus, 215 S. Ferry Rd, Narragansett, RI 02882, USA

[‖]Coastal and Hydraulics Laboratory, Engineer Research and
Development Center, U.S. Army Corps of Engineers,
3909 Halls Ferry Rd, Vicksburg, MS 39180, USA

[**]shoagland@vt.edu

[††]catherinej@vt.edu

‡‡*jirish@vt.edu*

§§*weiszr@vt.edu*

¶¶*kyle.mandli@columbia.edu*

‖‖*svitousek@gmail.com*

****catherine_johnson@nps.gov*

†††*mary.a.cialone@usace.army.mil*

As scientific understanding of barrier morphodynamics has improved, so has the ability to reproduce observed phenomena and predict future barrier states using mathematical models. In order to use existing models effectively and improve them, it is important to understand the current state of morphodynamic modeling and the progress that has been made in the field. This manuscript offers a review of the literature regarding advancements in morphodynamic modeling of coastal barrier systems and summarizes current modeling abilities and limitations. Broadly, this review covers both event-scale and long-term morphodynamics. Each of these sections begins with an overview of commonly modeled phenomena and processes, followed by a review of modeling developments. After summarizing the advancements toward the stated modeling goals, we identify research gaps and suggestions for future research under the broad categories of improving our abilities to acquire and access data, furthering our scientific understanding of relevant processes, and advancing our modeling frameworks and approaches.

1. Introduction

Coastal barriers are narrow landforms that are separated from the continental mainland by a shallow waterbody (Fig. 1). These barriers can be book-ended by inlets (i.e., barrier islands) or they can be connected to the mainland at one end (i.e., barrier spits) or both (baymouth barriers). The combination of backbarrier environment, subaerial island, and shoreface is often succinctly referred to as the "barrier system" or simply "barrier." As of 2011, over 20,000 km of the world's coasts were characterized by a

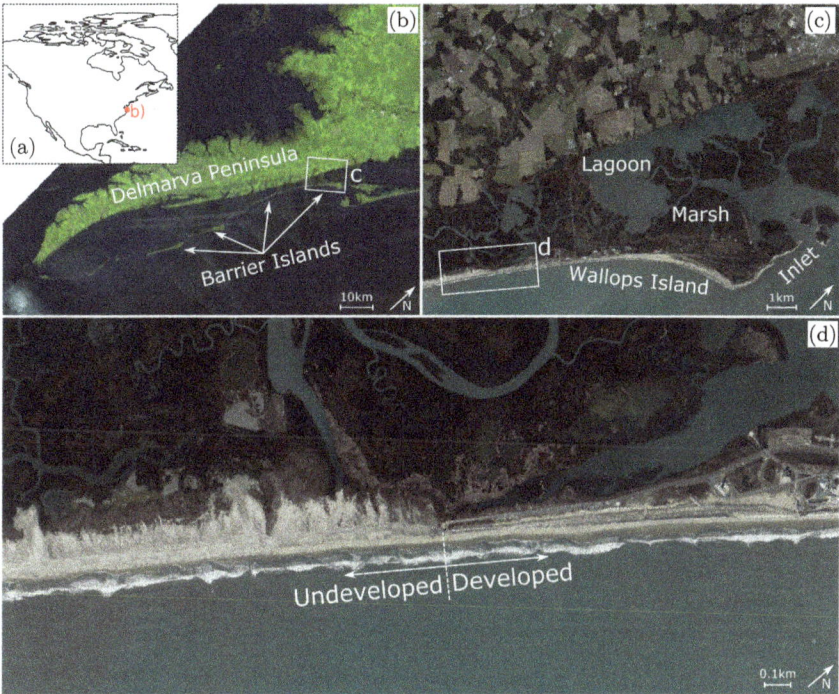

Fig. 1. Satellite and aerial images of a Virginia Barrier Island: (a) Location map. (b) Delmarva Peninsula (ESA, 2021). (c) Wallops Island (VGIN, 2021). (d) Zoomed section of Wallops Island (VGIN, 2021).

barrier system, accounting for approximately 10% of all coastlines (Stutz and Pilkey, 2011). Barriers provide significant benefits during coastal storms, such as surge volume and wave energy reduction, (Grzegorzewski *et al.*, 2011) wetland protection, (Wamsley *et al.*, 2009) sediment stabilization through the presence of subaerial or backbarrier vegetation, and protection of aquatic habitat (Bridges *et al.*, 2013). Additionally, barrier islands have become popular as both vacation destinations (Pilkey *et al.*, 2011) and permanent residential areas, which has led to increases in population density (Zhang and Leatherman, 2011).

Although many barriers have undergone rapid urban development since the mid-20th century, (Dolan and Lins, 1986) Stutz and Pilkey (2011) described this development boom as being "ironically" timed due to coastal hazard accelerations associated with current trends in sea level rise (SLR). According to the Intergovernmental Panel on Climate Change, global mean sea level (MSL) is predicted to rise between 0.25 and 1.0 m by the end of the century (Oppenheimer *et al.*, 2019). If these predictions hold true, the rates of barrier island morphological change and associated flooding during storms and other events will most certainly increase, (e.g., Gutierrez *et al.*, 2007). In addition to exacerbating coastal flooding, SLR also drives the evolution of the barrier system itself, influencing processes that change both the island's shape and location. Thus, on many barrier coastlines, permanent structures have been constructed on land that was and is expected to continue migrating toward the mainland.[5] Changes in the location and geometrical configuration of barrier systems are expected to alter the benefits that they provide to neighboring mainland communities. Therefore, it is critically important for all who are involved in coastal management to understand barrier island morphodynamics to produce the best possible outcomes for coastal communities.

While the earliest literature tended to document observations and initial theories of barrier morphodynamics, research has recently — in the last three or so decades — shifted toward the development and intensified use of computational models. Based on this observation, we note that where modeling often lagged behind or paralleled our advancements in scientific understanding, it has recently been used to validate and advance it. Many models have been developed over the last three to four decades. A review of these models may help new or future researchers survey the field of barrier morphodynamic modeling.

A few notable review papers have recently been published related to barrier morphodynamics. Some of these papers focus on a single, specific component of coastal change such as overwash, (e.g., Donnelly *et al.*, 2006), or storm sequencing and recovery, (e.g., Eichentopf *et al.*, 2019). Other reviews capture the larger-scale morphological response of barrier systems, but their application is either constrained to a particular location, (e.g., Rosati and Stone, 2009), or focused on a particular driver such as climate change, (e.g., Toimil *et al.*, 2020), or focused in-depth on a particular spatiotemporal scale, (e.g., Sherwood *et al.*, 2022). Table 1 provides a summary of these reviews and their focus areas. These reviews provide a valuable synthesis of relevant work but are not sufficient to capture the trends and advancements in barrier morphodynamic modeling.

The purpose of this manuscript is to fill that gap by providing a review of the literature regarding advancements in morphodynamic modeling of coastal barrier systems. Our review of modeling advancements is divided into two broad categories: (1) event-scale morphodynamics and (2) long-term morphodynamics — refer to the *Terminology* section for definitions

Table 1. Recent reviews.

Citation	Focus
Donnelly *et al.* (2006)	Laboratory work, field studies, and modeling efforts related to coastal overwash.
Rosati and Stone (2009)	Barrier evolution concepts from early literature; recent concepts in Northern Gulf of Mexico.
McBride *et al.* (2013)	Observations and conceptual models of barrier morphodynamics for various coastlines and regional locations.
Chardón-Maldonado *et al.* (2016)	Recent advancements on hydrodynamics and sediment transport modeling in the swash zone.
Reeve *et al.* (2016)	Long-term morphodynamic models that employ data-driven and/or hybrid approaches.
Ciavola and Coco (2017)	Event-scale processes and their impact on specific coasts (e.g., sandy beaches, barrier islands, tidal flats, etc.).
Moore and Murray (2018)	Compilation of recent work and synthesis of current understanding and state of research on barrier morphodynamics.
Eichentopf *et al.* (2019)	Laboratory studies, field work, and modeling exercises related to storm sequencing and beach recovery.
Ranasinghe (2020)	Commonly used morphodynamic models for sandy beaches and ideas for future long-term models.
Toimil *et al.* (2020)	Coastal erosion modeling, climate change impacts, and approaches for evaluating uncertainty.
Sherwood *et al.* (2022)	Advances in modeling event-driven morphodynamics on sandy coasts.

of "event-scale" and "long-term." These sections begin with a brief description of commonly modeled phenomena and processes, followed by a review of relevant modeling efforts, which are categorized according to their primary intent. At the conclusion of these sections, we summarize the primary contributions of the modeling developments and their limitations. Finally, we conclude with the identification of research gaps that currently exist and suggest directions for future research.

A few items should be noted regarding this study. First, there are some relevant topics (e.g., anthropogenic impacts and influences of vegetation) which are only briefly discussed due to our focus on morphodynamic modeling. Second, we have intentionally included many models and/or modeling approaches from the early literature so that the current models might be understood in their proper historical context, which requires knowledge of both previous and ongoing efforts. Additionally, this review primarily focuses on models in wide use in the research community. Therefore, some commonly used propriety models have only been briefly mentioned. Lastly, although our review is focused on barrier morphodynamics, many of the relevant processes play an important role for non-barrier coasts. Therefore, in order to fully understand the modeling advancements relevant to barrier systems, we must consider some modeling efforts that are not barrier-specific.

Before starting this review, it may be helpful to orient the unfamiliar reader by defining our modeling goals and our terms. In the following section, we have attempted to summarize our modeling goals with one overarching statement or *Grand Challenge*. This is followed by a brief discussion of terminology used in this manuscript.

1.1. *Grand challenge*

In theory, the ideal morphodynamic model would produce accurate predictions in a reasonable time without significant computational expense. As we consider how these ideals translate into reality, there are multiple modeling goals that we must work toward and important intermediate steps that we must first achieve. However, rather than outlining each goal, we have attempted to synthesize them into a single overarching goal, or *Grand Challenge*, as follows:

> *To predict barrier system morphodynamics in multiple spatiotemporal dimensions (e.g., short to long time scales, transect to regional evolution) with a high degree of confidence, under reasonable computational resources constraints, and considering relevant factors such as event-driven morphological change, evolution during non-stormy periods, biological processes (and other potential subsystem influences), and anthropogenic impacts.*

We intend the phrase "predicting... with a high degree of confidence" to mean predictions that have at least been partially validated and are useful in planning and decision-making. Throughout this review, the reader is encouraged to consider each development in light of the *Grand Challenge*. At the conclusion of each major section, we summarize the modeling advancements and extant limitations, offering our perspectives on progress toward this overarching goal. To maintain this focus, it should be noted that some relevant topics such as biological processes and anthropogenic impacts are given more of a cursory discussion.

1.2. *Terminology*

There are some inconsistencies in terminology in the body of work on barrier morphodynamics. Thus, for the purpose of this review, our aim here is to define terms that describe what is being modeled (e.g., a phenomenon or a process), the types of mathematical representations that are used (e.g., a model or a formulation), and the spatiotemporal scales used throughout the paper.

When discussing phenomena, we are talking about observable characteristics, behaviors, or events of a system. While the spatiotemporal scales of a system may vary (e.g., initiation of particle movement vs. shoreline behavior), there are phenomena associated with each system that may be mathematically represented via the development of a model. When we discuss processes, we are referring to patterned events that systematically contribute to the observable phenomena of a system. Based on these terms, we also distinguish between models and formulations. Whereas models are developed to represent phenomena, specific formulations are developed to represent processes. Models, therefore, may contain one or more formulations of a process. For example, consider the development and growth of a spit. The spit development and/or growth would be the observed phenomenon that is systematically progressed by the process of longshore sediment transport (LST). Thus, we might develop a model that predicts spit development and growth using a specific LST formulation.

The last terms that need to be defined up front are related to the spatial and temporal scales at which the relevant processes are typically resolved in coastal morphodynamic modeling. Herein, we adopt the temporal scale classification of Rosati and Stone (2009) and adopt a slightly modified version of the spatial scale classification of Cowell *et al.* (2003). These scales are presented in Table 2 and are used throughout this chapter. Note that we also use the term "event scale" throughout this book to refer to the combination of small spatial and short temporal scales.

Table 2.   Spatial and temporal scales of barrier island morphodynamics, respectively, modified from the works of Cowell *et al.* (2003) and Rosati and Stone (2009).

Type	Term	Scale
Spatial	Small scale	10^0–10^2 m
Spatial	Moderate scale	10^2–10^3 m
Spatial	Large scale	>10^3 m
Temporal	Short term	Hours to days
Temporal	Mid term	Days to decades
Temporal	Long term	Decades to centuries

2. Event-Scale Morphodynamics

This section provides an overview of commonly modeled phenomena and processes associated with event-scale morphodynamics, a review of relevant modeling efforts, and a summary of advancements toward the *Grand Challenge*.

2.1. *Commonly modeled phenomena and processes*

Acute sediment transport processes, which are characterized by a sudden onset and short-term duration, are initiated when a storm approaches the coast. Chronic transport processes, which are characterized by gradual beginnings and mid- to long-term duration, are not initiated during storms but are intensified by them. As these transport processes are initiated or intensified, the barrier responds in the form of morphological adjustment. To frame the discussion on storm response, we use the storm impact scale published by Sallenger (2000) wherein acute processes occur within four regime classifications: swash, collision, overwash, and inundation (Fig. 2). Each regime has certain morphological responses associated with runup levels.

In the *swash* and *collision* regimes, increased water levels by storm surge and wave runup lead to increased erosion on the beach and dune, depositing the eroded material seaward of the beach. *Collision* differs from *swash* in that the water level exceeds the dune toe, allowing waves to collide with and erode the lower parts of the dune slope, which can lead to avalanching of the upper dune. Sallenger (2000) points out that while

Runup/Surge Condition	Impact Regime	Associated Morphological Response
$R_{High} < D_{Low}$	Swash	Profile Erosion
$R_{High} > D_{Low}$	Collision	Dune Avalanching
$R_{High} > D_{High}$	Overwash	Erosion & Aggradation
$R_{Low} > D_{High}$	Inundation	Ocean-side Breaching
From Sallenger (2000)		

Runup/Surge Condition	Impact Regime	Associated Morphological Response
$B_{High} > D_{High}$	Outwash	Bay-side Breaching

Following terminology proposed by Over et al. (2021)

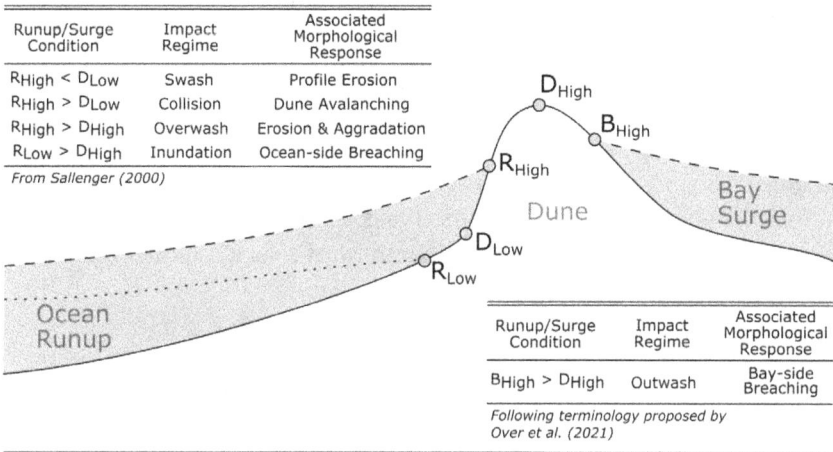

Fig. 2. Storm impact scale.

Source: Figure modified from the work of Sallenger (2000) with outwash regime.

sediment transported offshore under this regime may return to the beach, this sediment typically does not make it back to the dune structure, resulting in net erosion of the dunes over time. In the *overwash* regime, water levels are high enough such that incident wave runup intermittently flows over the dune peaks or antecedent low spots, carrying mobilized sediment with it. Lastly, the *inundation* regime involves complete submergence of the barrier which can lead to inlet formation (i.e., breaching) and significant increases in the cross-shore sediment transport (XST) rates (Sallenger, 2000). Inundation is associated with extreme levels of erosion that pick up normally dry (subaerial) sediment.

One regime that Sallenger does not include is the *outwash* regime, following the terminology proposed by Over *et al.* (2021) which describes seaward flows and associated offshore sediment transport. Although it is possible to have initial seaward surge depending on the orientation of the islands and the approach angle of the storm, initial surge levels are typically directed onshore. Therefore, seaward flows associated with the *outwash* regime usually occur after storms make landfall or pass by the area of interest, which reverses the predominant wind direction. Applied to a typical barrier system, this reversed wind field can cause backbarrier water levels to surge above receding ocean-side water levels. In this instance, breaching may be initiated from the backbarrier by outwash flows that scour a new channel through the island, liquefaction of previously weakened dune structures, or a combination of both. Various studies including (Over *et al.*, 2021; Shin, 1996; McCall *et al.*, 2010; Smallegan and Irish, 2017; Harter and Figlus, 2017) highlight the importance of considering this regime when modeling storm event morphodynamics.

The following sections offer an introductory discussion on commonly modeled phenomena and processes associated with barrier response to storm events, namely profile erosion and shoreface response, overwash, and breaching. This is followed by a review of relevant modeling efforts.

2.1.1. *Beach profile erosion and shoreface response*

While the term "profile" can be used to describe a wide range of the barrier system, we use the term "beach profile" herein to describe the morphodynamic response of the barrier's beach–dune complex and upper shoreface, which we loosely define as the morphologically "active zone" following Stive and de Vriend (1995) and Cowell *et al.* (2003). Generally, there are two primary factors that contribute to erosion of the beach profile under storm conditions:

(1) increased offshore-directed currents and (2) increased total runup. As the waves intensify, the beach profile state turns erosional (assuming a prior accretive state) as wave-driven sediment transport becomes dominated by undertow and rip currents which are offshore-directed (Aagaard and Kroon, 2017). Sediment is eroded from the upper portions of the profile and deposited on the shoreface, typically in a subaqueous bar, which is then delivered back to the profile once storm conditions subside (Quartel *et al.*, 2007). This cycle of erosion and subsequent recovery has been observed over seasonal wave-climate changes (Shephard, 1950) and event-scale changes (Ranasinghe *et al.*, 2012). Second, the total runup, as produced by a combination of storm surge, astronomical tide, and wave runup, may exceed the *swash* regime water level to collide with the dune and cause notching (i.e., erosion and recession of the lower dune), followed by slumping or avalanching (Edelman, 1968; Roelvink *et al.*, 2009). For a more thorough review of sediment transport processes during storms and relevant factors, including the role of infragravity waves and incident wave nonlinearity, the reader is referred to the work of Aagaard and Kroon (2017) and references therein.

These two primary factors (i.e., increased offshore-directed currents and increased total runup) contribute to barrier morphodynamics in significant ways. For example, in the *collision* regime, they lead to a net loss of sediment offshore to the lower (inactive) profile (Sallenger, 2000). This net loss effectively limits the ability of the beach and dunes to fully recover to pre-storm conditions without requiring external sediment sources (i.e., from the shelf, erodable profile outcrops, or LST gradients). Moreover, although much of the eroded sediment is brought back to the beach and dunes after the storm, this natural renourishment of the profile is not instantaneous but can take days or weeks to recover, (e.g., Quartel *et al.*, 2007; List *et al.*, 2004), leaving the barrier system in a temporarily hyper-vulnerable state. Profile recovery between storm events, although less studied than erosional events, is critically important to understanding barrier vulnerabilities to storm sequences and long-term morphology (Eichentopf *et al.*, 2019).

2.1.2. *Overwash*

Overwash occurs when water flows over the dunes. Sediment is carried by the water and deposited behind the dunes as washover. While overwash was associated with intermittent overtopping in Sallenger's *overwash* regime, it should be noted that by definition, overwash also occurs during Sallenger's *inundation* regime and the proposed *outwash* regime, as the landward or seaward directed flows continue to transport sediment across the dunes. Donnelly *et al.* (2006) offered distinct definitions for runup overwash and inundation overwash and discussed the differences and implications of each process.

Three factors are the primary contributors to increased likelihood of barrier island overwash: (1) antecedent low spots in barrier topography, (2) high water levels driven primarily by storm surge, and (3) large incident waves. Although it can be argued that this is self-evidently true, it is also confirmed in the early literature on barrier island storm response, (e.g., McCann, 1979; Cleary and Hosier, 1979). In addition to these three main contributing factors, overwash occurrence has also been associated with other variables including previous overwash activity, barrier island width, and vegetation density, (e.g., Cleary and Hosier, 1979; McCann, 1972; Fisher and Simpson, 1979). However, some of these factors can be indirectly related to antecedent topography. For example, areas that have experienced previous overwash events are also locations where the dunes have likely been lowered; thus, previous overwash activity can be linked to pre-storm discontinuities in the dune elevation. Similarly, since dune vegetation promotes sediment settling and dune growth, vegetation density could generally be considered a proxy for pre-storm topography. Donnelly *et al.* (2006) identified two other important factors including the direction of storm approach, which influences incident wave heights, and nearshore bathymetry, which impacts wave transformation.

Storms have significant morphological impact on barrier islands, which in turn affect the continued evolution and response to future storms. Observations from the early literature describe both destructive and constructive effects of overwash: destructive in that overwash may lower or destroy the dunes, (e.g., Nichols and Marston, 1939), and constructive in that overwash may contribute to aggradation of the barrier islands over time, (e.g., Rosen, 1979). Both of these effects directly impact flood risk from future storms. Again, to avoid duplicating work, the reader is referred to the review by Donnelly *et al.* (2006) which covers a variety of topics related to overwash including field and laboratory studies, modeling efforts, and its impact on barrier morphodynamics.

2.1.3. *Breaching*

Breaching is the creation of an inlet in a barrier that establishes direct hydraulic connectivity between the ocean and backbarrier water body.[44] Breaches have been shown to account for water level increases both during the storm event in the form of bay surge, (e.g., Cañizares and Irish, 2008), and after the storm event in the form of increased tidal range, (e.g., Conley, 1999). Excess flooding, property damage, habitat loss, and decreased navigability are possible negative outcomes from a breach; however, breaching is also desirable in some cases and may be intentionally performed in order to increase habitat connectivity for certain estuarine wildlife (Gerwing *et al.*, 2020) or to prevent undesirable back-barrier conditions, including low salinity, poor water quality, and, in some cases, flooding (Kraus and Wamsley, 2003).

From some of the earliest published observations of breaching, we know that multiple breaches, of various widths and depths, may form and expand during a single storm event, (e.g., Nichols and Marston, 1939). More recent studies have highlighted the dynamic nature of breaches, which can significantly change dimensions over relatively short time periods and even migrate alongshore, (e.g., Kraus and Wamsley, 2003; Wamsley and Kraus, 2005). Timing of the initial breaching process has received relatively little attention in the literature due to the difficult nature of collecting field data. However, a study by Visser (1998) and a related modeling exercise by Roelvink *et al.* (2009) estimated lateral growth rates of breaches between 1 and 2 cm/s during initial formation. During the phases of breach growth, XST is much greater than LST; however, once flow in the breach ceases, LST may cause closure of the breach (Kraus *et al.*, 2002).

In exploring the causes of breaching, researchers have often wanted to know on which side of the barrier breaching is initiated (Johnson, 1919). Multiple theories of breach formation are present in the early literature, as reviewed by Pierce (1970) including breaching from the backbarrier side through the escape of impounded water (Shaler, 1985) and ocean-side breaching by wave-driven erosion (Johnson, 1919). Pierce (1970) determined that barriers are most likely to breach from the lagoon side but stated that a narrow barrier could also be breached by erosion from the sea. Although this perspective was published as early as 1970, it received little attention until recent years, (e.g., McCall *et al.*, 2010; Smallegan and Irish, 2017; Harter and Figlus, 2017; Sherwood *et al.*, 2014).

Kraus *et al.* (2002) described two breaching processes and their association with lagoon-side or ocean-side breaching. The two processes are (1) scouring and channelization and (2) seepage and liquefaction. Scouring and channelization most commonly occur from the seaward side of the barrier, when sustained storm surge allows for water to (semi)continuously inundate the island with flow over the barrier; conversely, seepage and liquefaction typically initiate breaching from the landward side of a narrow barrier (Kraus *et al.*, 2002). However, recent modeling studies, (e.g., Shin, 1996; Smallegan and Irish, 2017; Sherwood *et al.*, 2014), have also shown that seaward-sloping water level gradients that occur after the ocean-side's peak storm surge have the potential to scour channels across the barrier as well that can lead to seaward sediment transport and breaching.

2.2. *Modeling efforts*

As stated previously, modeling efforts are classified according to their primary intent. Most event-scale morphodynamic models or formulations were developed to simulate a few key phenomena or processes including (1) beach and dune erosion, (2) shoreface response, (3) overwash, (4) breaching, and (5) combinations of categories 1 through 4. The following sections review the relevant modeling efforts which are also graphically summarized in Fig. 3.

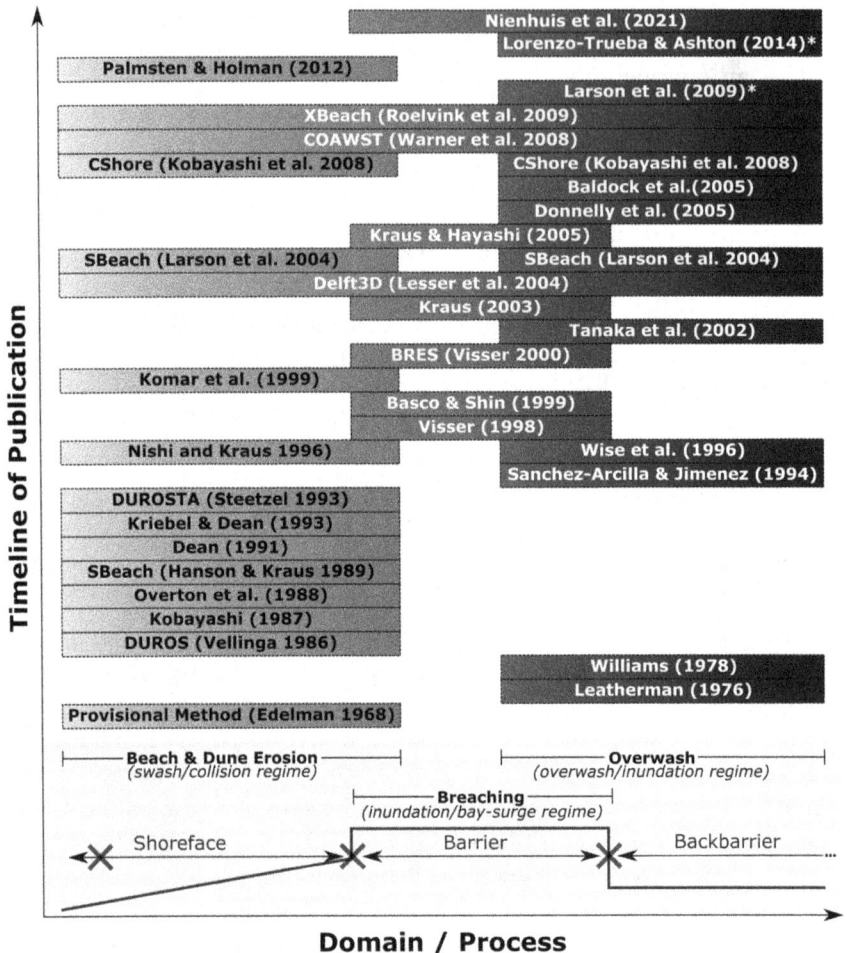

Fig. 3.   Event-scale models and formulations. Models are shown according to their publication chronology and are aligned with their respective processes, which range from beach and dune erosion, to breaching, to overwash.

2.2.1. *Modeling beach and dune erosion*

Modeling work on storm-driven response of the beach–dune complex was initiated and significantly advanced by researchers in the Netherlands in the 1960s and 1970s. Edelman (1968) observed that when storm surge levels exceeded the dune toe, the dune would undergo significant erosion and partial avalanching. Based on these observations, Edelman published

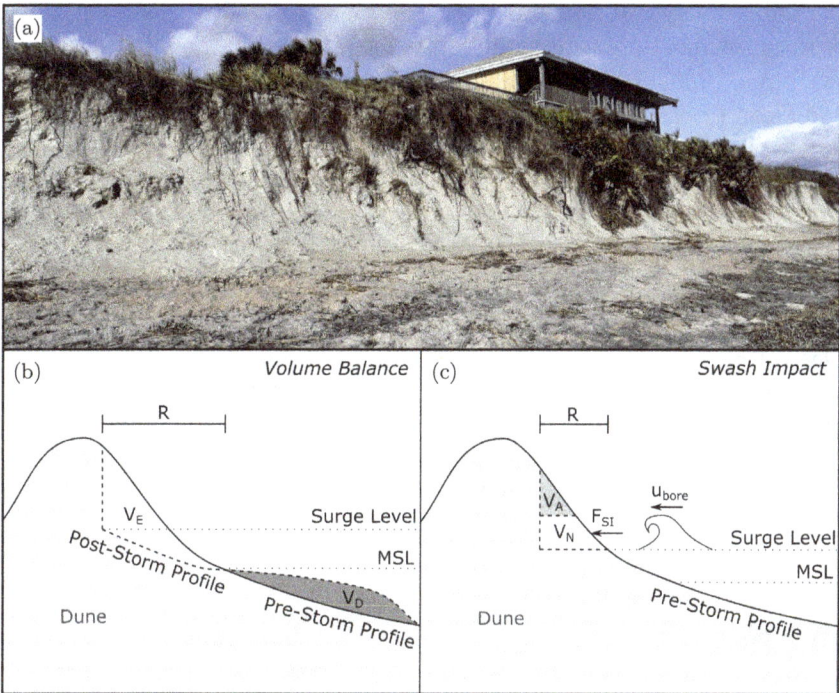

Fig. 4. Beach and dune erosion: (a) Image of beach and dune erosion from Hurricane Matthew (Brennan, 2016). (b) Volume balance approach that predicts dune recession (R) by equating the erosion volume (V_E) and deposition volume (V_D), modified from Edelman (1972). (c) Swash impact approach that relates wave bore velocity (u_{bore}) to the swash impact force (F_{SI}) which creates notching (V_N) that leads to avalanching (V_A), modified from the work of Nishi and Kraus (1996).

the first analytical formulation (i.e., method with a closed-form solution) for predicting dune erosion and retreat, later termed the "Provisional Method." This method assumed the formation of a new dune toe at the peak storm surge elevation and balanced the volume of sediment eroded from the dunes with deposition in the nearshore zone (Fig. 4(b)) using linear approximations of both the nearshore and dune profiles. Four years later, Edelman used the same principles to publish a similar method which considered more realistic (e.g., nonlinear) profile shapes (Edelman, 1972). In addition to sediment conservation and the new dune toe location, Edelman's work was based on other key assumptions including a constant profile shape, rapid (or instantaneous) profile response, and the presence of both storm and pre/post-storm equilibrium profiles.

Other analytical methods were developed to predict beach and dune erosion using similar assumptions; these models included DUROS (Vellinga, 1986) and those of Kobayashi (1987), Dean (1991), and Kriebel and Dean (1993). Fundamentally, each of these models is similar in that they are based on balancing eroded and deposited sediment volumes, while the main differences lie in the factors that influence the new profile shape. For example, the profile depth in the nonlinear Provisional Method was considered only a function of distance from the shoreline, (Edelman, 1972) while other methods allowed the depth to adjust based on factors such as wave height and sediment characteristics, (e.g., Vellinga, 1986; Dean, 1991). Komar *et al.* (1999) also developed a simple method to predict dune retreat based on the foreshore slope and the height of the runup above the dune toe; this approach was recommended by FEMA for United States Pacific Coast beaches as of 2005 (Mull and Ruggiero, 2014). Vellinga's DUROS model (Vellinga, 1986) continues to be used in the Netherlands to assess the health and safety of the coastal dunes (Bosboom and Stive, 2021).

One important limitation with these early models arises from the assumption of instantaneous response. Since the duration of a storm is often much shorter than the time required for profiles to erode to their new equilibrium states, they rather erode some fractional amount toward equilibrium but never reach it. Komar and Moore (1983) put it succinctly, stating that these methods *"should be regarded as an upper limit or an erosion potential that would result if the storm conditions were held constant indefinitely."* For conservative estimates and design standards, these methods may prove reliable. However, for higher levels of modeling accuracy, it may be necessary to shift toward time-dependent models or the combination of idealized models with a time-dependent function, (e.g., Komar and Moore, 1983).

Fisher and Overton (1984) proposed another type of modeling approach that focuses on the impact of swash on the dune face. These are appropriately called "Swash Impact" approaches. The main idea undergirding this approach is that erosion of the dune is proportional to the impact force of colliding waves, which can be related to the waves' bore heights and approach velocities (Fig. 4(c)). Through a series of laboratory experiments, linear relationships were found between the amount of dune erosion and swash impact force, modulated by statistically significant factors, such as grain size and dune density (Overton *et al.*, 1988, 1994). This relationship was also identified in the field through a series of experiments at Duck, North Carolina (Fisher, 1987).

Other methods using this approach were developed by Nishi and Kraus (1996); Larson *et al.* (2004) and most recently by Palmsten and Holman (2012). Nishi and Kraus (1996) calculated the swash impact force by multiplying the mass of water in the approaching wave by its deceleration upon impact. Using large-scale wave tank experiments on compacted and uncompacted dunes, they found linear relationships between the weight of eroded sediment and the impact force. They also found uncompacted sediment to be more susceptible to swash impact erosion and suggested artificially compacted dunes as a possible method of erosion control. Using the linear relationship between erosion and swash impact force as an initial assumption, Larson *et al.* (2004) derived an analytical model that predicted dune recession as a function of bore speed, initial geometry, empirical transport coefficients, and foreshore slope, which was assumed to linearly continue landward of the dune toe. The authors used four previously published datasets to test their model and to empirically derive an optimal transport coefficient. Lastly, Palmsten and Holman (2012) improved on this formulation in two main ways: (1) They used a Gaussian distribution to model variability in wave runup elevations and (2) they tested various runup exceedance values and found that using a runup exceedance value of 16% led to better dune erosion predictions in the laboratory when compared to the 2% runup exceedance guidance recommended by Sallenger (2000).

2.2.2. *Modeling shoreface response*

Paralleling these advancements was the development of more complex sediment transport formulations. While these formulations may vary in approach, they are similar in that they relate hydrodynamic parameters (e.g., velocity) to sediment transport rates. Thus, for the purposes of this discussion, we refer to these more complex formulations as coupled hydrodynamics–sediment transport (*HD-ST*) formulations. Since a review of each formulation would take considerable space, we offer a cursory description of the *HD-ST* formulations and refer interested readers to the works of Larson and Kraus (1989); Dean and Dalrymple (2002); Nielsen (2009); Aagaard and Hughes (2013) and Bosboom and Stive (2021) and references therein for additional details.

In highly resolved models, coupled *HD-ST* formulations use hydrodynamic parameters to predict both bed load and suspended load transport rates. Bed load transport is typically estimated as a function of the bed shear stress, sediment density, and average grain diameter (often using

Shields parameter), whereas the suspended sediment transport rate is calculated by integrating the vertical velocity and concentration profiles, the latter of which can be based on functions such as the Rouse profile or advection–diffusion calculations (Bosboom and Stive, 2021).

Depending on the application, not all models can afford the computational burden associated with coupled *HD-ST* formulations. Other approaches with less computational burden have gained popularity, such as the equilibrium-based approach, originally developed by Kriebel and Dean (1985) which assumes the existence of an equilibrium shoreface profile that controls how the shoreface responds under specific hydrodynamic conditions. It is founded on the idea that if hydrodynamic conditions remained constant, then the shoreface would respond until constructive (landward) and destructive (seaward) forces along the profile were balanced, leading to a steady profile with an XST rate of zero. Kriebel and Dean (1985) developed a formulation that calculates an equilibrium profile based on depth-dependent energy dissipation rates. XST rates are then calculated at a particular shoreface depth based on the difference between the actual energy dissipation rate and the rate associated with the equilibrium profile (Dean and Dalrymple, 2002).

Another popular approach is the energetics approach, which was originally developed by Bagnold (1963) for fluvial sediment transport. This approach considers the hydrodynamic environment as a machine that performs a certain amount of work (sediment transport) based on the available power input (kinetic energy) modulated by some efficiency factor (resistance to transport) (Bagnold, 1966). Bed load and suspended load transport rates are calculated separately based on the available wave power, or the wave energy flux per unit width, which drives the transport (Dean and Dalrymple, 2002). While the energetics approach has been successful in predicting offshore-directed sediment transport rates during storm events, this approach has generally underpredicted onshore sediment transport during recovery periods (Aagaard and Hughes, 2013).

2.2.3. *Modeling overwash*

Efforts to quantitatively understand and predict overwash have led to the development of various formulations, which generally fall into one of two categories. Those in the first category may be described as "Bulk" approaches, as defined by Donnelly *et al.* (2006) since they relate certain hydro-dynamic parameters (e.g., wave height) to bulk washover volumes

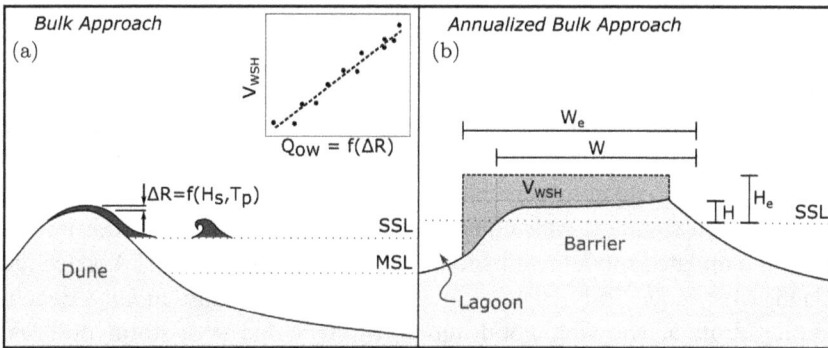

Fig. 5. Overwash modeling approaches: (a) Traditional bulk approach that predicts washover volume (V_{WSH}) based on bulk parameters (e.g., excess runup height ΔR), modified from the work of Donnelly *et al.* (2009). (b) Annualized bulk approach that predicts V_{WSH} based on width (W) and height (H) deviations from equilibrium values (W_e and He) based on the storm surge level (SSL), modified from the work of Lorenzo-Trueba and Ashton (2014).

(Fig. 5(a)). Williams (1978) published the earliest bulk, which predicted the washover rate as a function of excess runup (i.e., depth of runup over the dune crest) and wave period. Later bulk formulations, (e.g., Tanaka, 2002), were based on laboratory experiments by Kobayashi *et al.* (1996) which showed a linear relationship between overwash and washover rates. Formulations in the second category apply coupled *HD-ST* formulations, which were discussed in the previous section. Donnelly *et al.* (2006) reviewed at least three of these formulations and their results including Leatherman (1976) who coupled the Einstein transport equation to velocity measurements, Sánchez-Arcilla and Jiménez (1994) who combined the Van Rijn formulation with velocities calculated using the Chezy equation, and Baldock *et al.* (2005) who applied a standard sheet flow model based on Shield's parameter to calculated swash velocities.

In the last 15 years, most overwash modeling efforts have been directed toward developing, improving, and applying the coupled *HD-ST* formulations, which are typically just one component of event-scale morphodynamic models that resolve multiple sediment transport processes at small spatial scales. At the time of Donnelly's (2006) review, only one such model (i.e., SBEACH) was able to predict overwash. The original formulation, developed by Wise *et al.* (1996) predicted sediment transport landward of the swash zone boundary based on the estimated wave bore velocity at the dune crest and interpolated the transport rate to both landward and seaward limits. This formulation was later updated by Larson *et al.* (1976) who modified landward flow dissipation by including a lateral spreading component and Donnelly *et al.* (2005, 2009) who used the Sallenger (2000) regimes to model intermittent overwash by wave runup and quasi-steady overwash during barrier inundation, the latter of which used a standard weir equation. Donnelly *et al.* (2005, 2009) compared the updated model results to post-Hurricane field data at Assateague Island, Maryland, Folly Beach, South Carolina, and Garden City Beach, South Carolina, showing good agreement with the post-storm profiles. Additionally, Donnelly's model was shown to outperform that of Larson *et al.* (2004) in predicting the post-storm profile at Assateague.

Recent work has also involved the incorporation of bulk overwash formulations into long-term and large-scale barrier evolution models. The long-term model of Jiménez and Sánchez-Arcilla (2004) employs a bulk formulation for modeling overwash rates based on empirically derived annual overwash volumes. This formulation uses the critical length concept of Leatherman (1979) which posits a theoretical threshold (i.e., critical barrier width and height) at which overwash is prevented. Deviations from these critical thresholds are used to estimate accommodation space (or volume) in the subaerial and backbarrier zones (Fig. 5(b)). Thus, event-driven overwash is modeled continuously and quantified by the available accommodation space up to some predetermined maximum annual overwash volume. More recent models, (e.g., Lorenzo-Trueba and Ashton; Lorenzo-Trueba and Mariotti, 2017), also use the critical length concept to model overwash in their long-term models.

Larson *et al.* (2009) followed a different approach, developing an analytical method to simulate the retreat of the barrier (or dune) based on landward (i.e., overwash) and seaward (i.e., profile erosion) sediment fluxes. Using a triangular approximation for the island or dune, these flux values were correlated with the ratio of dune crest to total runup elevations, and validation with field data showed results could provide order of magnitude estimations of overwash flux.

2.2.4. *Modeling breaching*

In modeling a breach, there are a number of important components that one may wish to consider including the location of breach occurrence, the timing of breach formation, breach dimensions and its progression (i.e., expansion or contraction), and finally its ultimate state (e.g., natural closure and stable inlet). While there has been some quantitative work on predicting systematic breach occurrence, (e.g., Kraus *et al.*, 2008), and long-term inlet stability (see the work of Kraus and Wamsley (2003) and references therein), our focus will be limited to models with strong morphodynamic components (i.e., breach formation, initial breach growth, and long-term progression).

Visser (1998) developed a conceptual model of breach formation and initial growth. Although the model was originally developed for sand dikes, it can also be applied to barrier islands. The conceptual model described five phases: (1) erosion and steepening of the inner slope of the scour channel, (2) decreasing of the crest width, (3) crest lowering and breach widening, (4) breach widening as flow changes from critical flow to subcritical flow, and (5) breach widening during subcritical flow until the flow ceases. This conceptual model was translated into BRES, a numerical model that predicts breach formation and initial growth based on discharge (calculated using the broad-crested weir equation) through an initial trapezoidal cross-section (Visser, 2001). Testing against multiple laboratory and field studies, Visser (2001) found good agreement between predicted breach widths over time and measured data.

Basco and Shin (1999) published a 1D numerical breaching model based on storm stages, in a similar fashion to Sallenger's regimes (Sallenger, 2000). Dune erosion was modeled in the first stage, followed by a diffusion-based approach to overwash in the second stage. The third stage aligns with Sallenger's *inundation* regime, while the fourth stage aligns with the *outwash* regime. In these last two regimes, barrier inundation and breaching were modeled by combining the 1D Saint-Venant equations with the sediment transport formulation of Van Rijn (1984).This approach to breach modeling has been included in more recent event-scale morphodynamic models (e.g., Delft3D and XBeach), which combine hydrodynamic output with specific sediment transport formulations. These models predict breach formation during barrier inundation, when flow velocities across the island scour antecedent low spots into fully formed channels. Additional details on these models may be found in the following section.

Kraus (2003) developed an analytical breaching model that predicts the development of a rectangular breach toward equilibrium dimensions using an exponential time function. The model starts with some initial channel or non-uniformity in the dune or island and proceeds toward a full breach based on flow through the channel which erodes the channel bed and sides. Kraus (2003) found the breach response to be sensitive to initial channel dimensions. Kraus and Hayashi later expanded the model to include a coupled *HD-ST* formulation, where breach progression was based on calculated bottom and critical shear stresses. The model was shown to reproduce general trends of an observed breach, yet it tended to underestimate the breach width and overestimate the breach depth (Kraus and Hayashi, 2005).

A more recent analytical breaching model was developed by Nienhuis *et al.* (2021) that is based on the hypothesis that a breach develops when the volume of sediment transport by overwash exceeds the sediment volume stored in the subaerial island. Overwash volume is calculated analytically using a triangular storm surge time series and integrating an overwash flux equation that considers surge height, width and depth of the dune gap, and a friction coefficient to account for vegetation impacts. Nienhuis *et al.* (2021) compared their model results to Delft3D simulations and found that it performed reasonably well, although the Delft3D predictions varied across one additional order of magnitude compared to the analytical model. Results were also compared with observations from Hurricane Sandy which showed that the model performed much better for undeveloped barriers as compared to developed barriers.

2.2.5. *Multifaceted event-scale modeling*

A variety of morphodynamic models have been developed to simulate more than one event-scale phenomena/process — we refer to these as "multi-faceted" models. Readers familiar with the literature will recognize that many of these multifaceted models are commonly called "process-based" models, although we have intentionally avoided this term due to its inconsistent and ambiguous usage in the literature, as well as its implication that more abstracted models are not based on processes. In the following, we present select event-scale models, followed by a brief discussion of multifaceted modeling efforts related to storm sequencing and post-storm recovery, which has received less attention from researchers until recently.

Event scale models
While a variety of multifaceted event-scale models exist, herein we focus on models that have been thoroughly cited in the literature and are widely used by the coastal morphodynamics research community. These include models such as SBEACH (1989), which rely on equilibrium concepts, and

models such as DUROSTA (also known as Unibest-DE), (Steetzel, 1993) CShore, (Kobayashi *et al.*, 2008) Delft3D, (Lesser *et al.*, 2004) and XBeach, (Roelvink *et al.*, 2009) which are based on coupled *HD-ST* formulations. Some of the primary differences between these models are shown in Table 3, including model dimensionality, included processes, and process formulations. In the following, we discuss the development of each model and highlight some significant improvements. Readers are referred to the references provided with each model for additional details.

SBEACH (1989) was developed in the late 1980s to predict profile response to storm events. The model employed the XST formula of Kriebel and Dean (1985) which is based on the difference in energy dissipation between the actual profile and an equilibrium profile. The model was originally calibrated using data from large wave tank experiments, showing its ability to predict foreshore erosion and bar formation, and its inability to predict features landward of the bar such as the trough and berm development during accretionary simulations (Larson and Kraus, 1989). The original model (which did not include overwash) was formally updated with the overwash formulations of Wise *et al.* (1996) and again by Larson *et al.* (2004) who showed good agreement between model predictions and measured profile changes for observations at Ocean City and Assateague, Maryland. SBEACH has more recently been incorporated in economic models for evaluating beach nourishment projects, (e.g., Gravens *et al.*, 2007), probabilistic frameworks for predicting erosion, (e.g., Callaghan *et al.*, 2013), and model comparison studies, where it produced better morphological predictions than XBeach when using default parameters but underperformed when calibration data were employed, (e.g., Callaghan *et al.*, 2013; Simmons *et al.*, 2019).

DUROSTA, which is an acronym in Dutch for "dune erosion — time dependent," was developed in the early 1990s as an unsteady, numerical model upgrade to the analytical beach and dune erosion models DUROS (Vellinga, 1986) and DUROS+ (the "+" representing the addition of wave period to the original model parameterization). The model was initially validated by comparison to laboratory data and various field experiments and showed good prediction capabilities on the subaqueous profile while underestimating dune retreat (Kobayashi *et al.*, 2008). DUROSTA was used by Van Baaren (2007) who found that wave period, bed slope, and the location of transition between the wet and dry profile zones were important model parameters. Hoonhout (2009) also used the DUROSTA model to study the effects of shoreline curvature on dune erosion and

Table 3. Multifaceted morphodynamic models.

Model Name	Refs.	Dim.	Process Formulations[†]				Model Description
			XST	LST	OW	BR	
SBEACH	(Larson and Kraus, 1989)	1D	KD85		WIS96		XST rates estimated through semi-empirical relationships in shoreface regions; considers wave and sediment characteristics, wave shoaling, breaking, setup and set-down, breaker decay and reformation, sediment slumping/avalanching.
DUROSTA/ Unibest-DE	(Steetzel, 1993)	1D/Q2D	[...STZL93...]				Only considers suspended load transport (bed load neglected); considers wave set-up, energy dissipation from bed friction after breaking with a turbulence model; employs a bed slope correction factor and extrapolates swash transport rates based on calculated rates at the wet/dry interface.
CShore/ C2Shore	(Kobayashi and Farhadzadeh, 2008; Grzegorzewski et al., 2013)	1D/2D	[.... KBY08....]		KBY10		Hydrodynamic components include the combined action of incident waves and currents, considering wave shoaling, breaking, and roller energy; considers shoreface (or structure) permeability and overtopping using an empirically based, probabilistic runup model.

| Delft3D | (Lesser et al., 2004) | 2D/3D | [............VRN93.............] | Shallow water equations solved in 2D (depth-averaged) or 3D; allows coupling to HISWA or SWAN wave models which consider breaking, bed friction, and streaming (near-bed currents); includes surface roller and infragravity formulations; includes bed slope correction and morphological acceleration factor. |
| XBeach | (Roelvink et al., 2009) | 2D | [............SVR97............] | Depth-averaged shallow water equations solved in Sallenger (2000) storm impact regimes; includes wave breaking, swash dynamics (modeling wave groups, infragravity waves, surface rollers, and return flows), beach and dune erosion (including avalanching), overwash (using low-frequency wave group forcing), and breaching by channel scouring. |

Notes: †OW: Overwash; BR: Breaching; KD85: Kriebel and Dean (1985), WIS96: Wise *et al.* (1996) STZL93: Steetzel (1993), KBY08: Kobayashi and Farhadzadeh (2008), KBY10: Kobayashi *et al.* (2010), VRN93: van Rijn (1993), SVR97: Soulsby-van Rijn (1997).

retreat during storm events, finding that consideration of shoreline curvature significantly impacted the model results. Currently, DUROSTA and another cross-shore model Unibest-TC (Ruessink *et al.*, 2007) are optional modules that may be employed when using the one-line model Unibest-CL+.

De Goede (2020) presented a historical review of the development of Delft3D, from initial 2D shallow water code development in the late 1960s, to coupling of updated wave models (e.g., SWAN), to the addition of turbulence closure models for 3D flows in the 1990s, and finally to the incorporation of sediment transport formulations into the hydrodynamic module. Lesser *et al.* (2004) presented details on the latter update, as well as the inclusion of a morphological acceleration factor for long-term simulations and validation studies showing that the results compared well to analytical solutions, laboratory data, and other accepted numerical model solutions. Delft3D is widely used in both practice and research, (De Goede, 2020) including studies on event-scale flooding, (e.g., Cañizares and Irish, 2008), storm sequence morphodynamics, (Cañizares and Irish, 2005), breach stability and growth, (e.g., Alfageme, 2007), and morphodynamic changes between storm events, (e.g., van Ormondt, 2020).

Johnson *et al.* (2012) presented a thorough summary of the historical development of CShore from its initial goals in modeling nonlinear wave transformation in the late 1990s, to aiding in coastal structure design, and finally to its development toward modeling nearshore morphodynamics in the late 2000s. Johnson *et al.* (2012) also provided results from sensitivity analyses, model calibration, and validation at nine field sites, which showed the model was capable of producing reasonable estimates of event-driven morphological changes while tending to under-predict dune erosion and retreat. Work and improvement on the model have continued through at least 2015, (Kobayashi, 2016) and the model has also been extended to two dimensions (C2Shore), the latter of which was validated through simulations of morphological response to Hurricane Katrina at Ship Island, Louisiana (Kobayashi, 2016). CShore does not explicitly model sheet flow or ebb currents, reducing its applicability during barrier inundation (Harter and Figlus, 2017).

XBeach is considered the state-of-the-art event-scale model to predict barrier response to storm events. Lead by Roelvink *et al.* (2009) XBeach was developed as an open source model to predict all of the main morphological responses associated with storm events (i.e., beach and dune erosion, overwash, and breaching) corresponding to the storm impact regimes of Sallenger (2000). Model validation studies showed it was able to predict storm hydrodynamics and morphological responses well, (Roelvink, 2009) although subsequent studies have shown that high simulated velocities in the swash zone consistently led to slight overpredictions of erosion near the dune toe, (e.g., Van Dongeren *et al.*, 2009; De Vet, 2014). To correct these overpredictions, researchers have attempted to artificially lower sediment mobilization (by modifying the critical Shield's number); however, while this led to more accurate predictions of dune toe erosion, it decreased the accuracy of breaching simulations (De Vet, 2014). Elsayed and Oumeraci (2017) found that modifying suspended sediment concentrations based on the local bed slope helped resolve this issue. Some of the most recent work with XBeach has involved modifying roughness coefficients. Passeri *et al.* (2018) implemented spatially varied roughness coefficients based on land cover, which showed improved morphodynamic predictions over simulations with constant roughness values. Alternatively, van der Lugt *et al.* (2019) implemented dynamic roughness values that vary during the simulation according to erosion and deposition patterns, which showed improved results over simulations with static roughness values.

Many of these event-scale models continue to be tested and applied today. Although XBeach has become the standard for modeling event-scale morphodynamics, recent comparison studies indicate that other models (e.g., CShore, SBEACH, and Delft3D) are also being used and evaluated for their strengths, (e.g., Harter and Figlus, 2017; Simmons *et al.*, 2019; Cho *et al.*, 2019). Furthermore, various studies have loosely coupled these event-scale models together to utilize the strengths of each model. For example, Cañizares and Irish (2008) used SBEACH to simulate dune erosion and lowering prior to inundation and breaching using Delft3D. XBeach and Delft3D have also been loosely coupled in a recent breaching study by van Ormondt *et al.* (2020) who used XBeach to simulate breach development during the storm and Delft3D to simulate breach development and growth after the storm event.

Model coupling has also been utilized in the development of new modeling systems. The COAWST modeling system, which was developed by coupling a regional ocean model (i.e., ROMS), a nearshore wave model (i.e., SWAN), and an open-source sediment transport model (i.e., CSTMS), (Warner et al., 2010) is appearing more frequently in the coastal morphodynamics literature, including specific application to shoreline change modeling, (e.g., Safak et al., 2017), and barrier islands, (e.g., Safak et al., 2016; Warner et al., 2018). Numerous other modeling systems have been developed (see Kaveh et al., 2019) but have yet to gain a literature foothold in this particular field of study.

Storm sequences and post-storm recovery
Some of these event-scale models have also been applied to the study of storm sequences, which investigates the nonlinear impact of sequential storms on beach and dune erosion, where successive smaller storms have a cumulative effect that exceeds the impact of an independent event (Senechal et al., 2017). Various modeling studies have been conducted to quantify this cumulative impact and to determine the most important driving factors, such as antecedent beach states, (e.g., Splinter et al., 2014), and the order of the most severe storms within the sequence, (e.g., Dissanayake et al., 2015).

Based on a survey of the literature, Eichentopf et al. (2019) identified three primary conceptual descriptions to aid in modeling the impact of storm sequences and discussed evidence from published studies for each description. The three conceptual descriptions are as follows: (1) initial storm destabilization, where the first storm in the sequence erodes the beach, leaving it more vulnerable to the next storm event, (2) extreme storm impact, where the largest storm event of the sequence is of primary importance regardless of storm order, and (3) benchmark storm impact, where all events in a storm sequence may be combined and modeled as a single large storm event, similar to a benchmark or design storm approach in hydrologic analysis. Various types of models have been employed and/ or developed to study storm sequences including statistical models, (e.g., Pender and Karunarathna, 2013), long-term equilibrium-based models such as ShoreFor (Davidson et al., 2017) or PCR (Ranasinghe et al., 2012) and multifaceted event-scale models such as XBeach and Delft3D, (e.g., Splinter, 2014; Dissanayake, 2015).

In addition to reviewing the literature on storm sequencing, Eichentopf et al. (2019) also provide a brief section on recovery, which they indicate is much less studied than the impact of storm sequences. They concluded with recommendations for future research, which broadly included additional physical and numerical simulations, improved data collection efforts, and stronger research emphasis on beach recovery processes.

2.3. *Summary of advancements and limitations*

The practice of modeling event-scale barrier morphodynamics has followed a natural progression from conceptualizing models based on observations, to the creation of simplified and efficient rule-based models, and to the development of more complex sediment transport formulations coupled with hydrodynamic calculations at fine spatiotemporal scales. Reconsidering our *Grand Challenge* statement, it is apparent that significant advancements have been made over the last fifty years. The earliest and most basic models (e.g., analytical dune erosion models) were intuitive, easy to use, and could provide conservative estimates for dune recession and likelihood of failure. Empirical studies followed, which advanced our ability to quantify the impact of key processes based on hydrodynamic output (e.g., predicting notching/avalanching of the dune face based on swash impact, predicting overwash volumes based on runup exceedance, and predicting sediment transport rates based on velocity and concentration profiles). This improvement in scientific understanding, along with the advancements in computing power, has allowed us to continue reducing the spatiotemporal scales of our morphological predictions while maintaining or increasing accuracy.

However, there are still major limitations to our modeling capabilities. Although the accuracy of simulations has improved, we are still a long way from high-confidence predictions. This is partially due to the scarcity of data to evaluate the predictive capability of models mid-storm. Event-scale models are able to capture the general trends of erosion and deposition compared to pre-and post-storm profile (or LiDAR) data; however, the small-scale predictive abilities of our models during storms are largely unknown since there is little to no data to validate those predictions. Our apparent distance from high-confidence predictions can also be attributed to both epistemic uncertainty (i.e., that which arises from our lack of knowledge of the relevant processes) and intrinsic uncertainty (i.e., that which arises from the inherent randomness of natural processes). For example, we know that some factors — such as vegetation and anthropogenic impacts — play an important role in event-scale morphodynamics, yet the modeling of such factors is (for various reasons) still in its infancy.

Additionally, the inherent randomness of forcing conditions (e.g., storm characteristics and wave climates) and initial conditions (e.g., bathymetry and sediment characteristics) is difficult to capture at smaller scales.

3. Long-Term Morphodynamics

This section provides an overview of commonly modeled phenomena and processes associated with long-term morphodynamics, a review of relevant modeling efforts, and a summary of advancements toward the *Grand Challenge*.

3.1. *Commonly modeled phenomena and processes*

During the periods of time in between storm events, chronic sediment transport processes resume their work that contributes to gradual morphological change. The following sections discuss commonly modeled long-term phenomena (i.e., shoreline change and barrier transgression) and relevant morphodynamic processes.

3.1.1. *Shoreline change*

The shoreline can be smoothed or caused to vary in form depending on the angle of the incident waves which drive LST (Ashton *et al.*, 2001). Thus, shoreline change is observed as the local shoreline is moved either landward or seaward by gradients in LST rates. These gradient-driven changes can also manifest themselves in other ways including island migration, barrier elongation, inlet migration, and island dimensional changes.

Although it is not as common, entire barrier islands can migrate in the direction of LST when sediment is eroded from the updrift end, carried alongshore, and deposited at the downdrift end, assuming no updrift sediment sources. Otvos Jr. (1970). noted this phenomenon in the northern Gulf of Mexico by observing that barriers can migrate large distances (i.e., several kilometers) from their location of origin. When the barriers are stable and not prone to migration, newly formed inlets may migrate instead. This phenomenon results from a LST gradient across the inlet, where sediment is deposited updrift of the inlet and eroded downdrift.

Dimensional changes may also be observed due to LST gradients and the placement of engineering structures. McCann (1979) observed that most islands developed greater widths on the downdrift end of the island as compared to the updrift end, which was attributed to a minimal amount of updrift sediment available for transport. If a continuous source of updrift sediment is present, and sediment is not removed from the barrier system, then barrier elongation could be observed as sediment is continually added to the downdrift end. Penland and Boyd (1981) described lateral migration of barrier islands and the influence of placing coastal structures at various locations along the islands. For example, structures placed near the updrift end tended to reduce the total island area while structures placed in the middle of the island tended to increase the total area.

3.1.2. *Barrier transgression*

In addition to shoreline change, most barrier islands are undergoing transgression (i.e., landward migration) in accordance with SLR. However, this migration did not appear to be widely accepted in some of the earliest literature, (e.g., Schwartz, 1973; Leatherman, 1987). Nevertheless, once transgression was recognized by the research community, many studies sought to identify the driving mechanisms that were primarily responsible for it. Otvos Jr. (1970) indicated that overwash and aeolian processes were primarily responsible for the landward movement, which was supported by others, such as Moody (1964), Leatherman (1987), and Godfrey (1970). Others found sediment transport through tidal inlets and/or breaches to play a much larger role, (e.g., Fisher and Simpson, 1979; Leatherman, 1979; Pierce, 1969; USACE, 1984).

SLR rate is also considered one of the primary drivers of barrier transgression through its interaction with storm processes, such as overwash and breaching. Although not developed specifically for barriers, the Bruun Rule (1962) exemplifies the theorized direct relationship between SLR and shoreline transgression. The interaction between rates of SLR and other transgressive processes was published in an interesting study by Moslow and Heron Jr. (1979). They found that previous high rates of SLR were correlated with dominating overwash processes and high rates of transgression. Conversely, when the rate of SLR slowed, they found that transgression also slowed and inlet dynamics became the dominant method of sediment transport between the ocean and backbarrier environment.

During landward transgression, barrier islands may also maintain their elevation with respect to SLR through the combination of overwash and inlet dynamics/breaching. As SLR effectively reduces barrier island relief, barriers are more prone to overwash and inundation during storm events, which deposit sediment on the island or behind it (i.e., washover deposits). This deposition effectively translates the island landward and increases its elevation. As this process is sustained, the barrier sediment may be conceptualized as "rolling" over itself, which has led to the description of this cycle as "barrier rollover (Moore and Murray, 2018)." Lorenzo-Trueba and Ashton (2014) referred to this sustainable behavior as dynamic equilibrium.

Similarly, lagoonal washover deposits and flood tidal shoals have been shown to assist the barrier in maintaining its elevation through the reduction of accommodation space for future washover (Stolper *et al.*, 2005). For example, consider a salt marsh that grows on top of washover deposited in a lagoon during some initial storm event. When a subsequent storm arrives, sediment that would have been deposited in the lagoon is now deposited on top of the new salt marsh. Thus, the salt marsh (and previous washover deposit) acts to reduce the available lagoon space for washover, and elevation is increased in that location as a result. Recent modeling work has suggested that the presence of backbarrier marsh not only increases island elevations but also actually reduces landward transport by encouraging the subaerial deposition of sediments (Johnson *et al.*, 2021). As the barrier continues its rollover toward the mainland, those previously buried marsh and lagoonal sediments may show up as shoreface outcrops which can affect the future morphodynamics through changes in the sediment supply (i.e., the source of sediment that feeds the growing barrier).

Although sustained barrier transgression is associated with increases in subaerial elevations with SLR, barriers may also lose elevation due to compaction of the underlying sediment. Hoyt (1969) was possibly the first to mention the idea of vertical movement by compaction or isostatic adjustment. He stated that *"compaction or isostatic movement caused by weight of the sediment deposited in the coastal area may result in formation of lakes or lagoons by depression of the chenier plain below water level."* As the barrier rolls over previous marsh sediment, the marsh sediment compacts under the load of the island, inducing an even higher local rate of SLR.

Barrier island transgression is also considered to be influenced by two other factors: (1) the slope of the shelf over which it is migrating and (2) the sediment supply. If we only consider the geometry of the system and assume that barriers maintain their dimensions, it is apparent that barriers must migrate at higher rates over shallower slopes in order to keep pace with SLR (Pilkey and Davis, 1987). Numerous studies have concluded that antecedent topography is extremely important to the development and configuration of modern day barrier islands, (e.g., Halsey, 1979; Oertel, 1979; Belknap and Kraft, 1985). Others have concluded that sediment supply is more important to the rate of migration, with less sediment supply leading to increased migration, (e.g., Swift, 1975; Storms *et al.*, 2002; Moore, 2007; Ruggiero, 2010). Dillon (1970) commented on the cross-shore migration of barriers through stratigraphy observations and concluded that barriers were not forced to continue landward migration with SLR but could drown if the sea level advanced too quickly or if there was an insufficient supply of sand.

3.2. *Modeling efforts*

Perhaps the most challenging question related to barrier morphology is as follows: "What will be the state of a barrier system 10, 100, or even 1000 years from now?" Compared to analyzing and predicting short-term responses, there is considerably less evidence available (that is, evidence or data collected using our current era's level of scientific certainty) to evaluate historical trends and make long-term projections. Stratigraphic observation and analysis may provide a partial glimpse of historical system states; however, it also requires assumptions and a hermeneutic to make the evidence meaningful, thereby reducing the certainty of conclusions that may be drawn. On the other hand, there are also problems when extrapolating small-scale processes to large spatiotemporal scales (i.e., the problem of error propagation). Thus, the problem of long-term morphological analysis and prediction is not a trivial one, especially since it is closely tied to uncertainties surrounding climate change (e.g., future SLR and changes in storminess). Numerous publications from the early 1990s into the early 2000s discuss the philosophy behind long-term morphological prediction. The interested reader is referred to (Stive *et al.*, 1990; Terwindt and Battjes, 1990; De Vriend, 1991; Latteux, 1995), for further details on this topic.

Similar to the previous section, the review of long-term morphodynamic modeling efforts is broken down according to the primary intent of each model. Thus, modeling efforts are categorized by those which model (1) shoreline change, (2) shoreface evolution, (3) barrier transgression, and (4) phenomena that are typically combinations of categories 1–3.

To assist the reader in keeping track of the models discussed, Fig. 6 offers a graphical representation of long-term models, in the chronology of their publication, that simulate some combination of shoreface evolution, shoreline change, dune growth/erosion, or overwash. Table 4 is a comprehensive summary of the long-term models discussed in this review, which includes each model's relevant processes.

3.2.1. *Modeling shoreline change*

Long-term modeling of shoreline change is often referred to as "shoreline evolution" modeling since the most observable impact of LST gradients is shoreline displacement, either landward or seaward, the first approach to

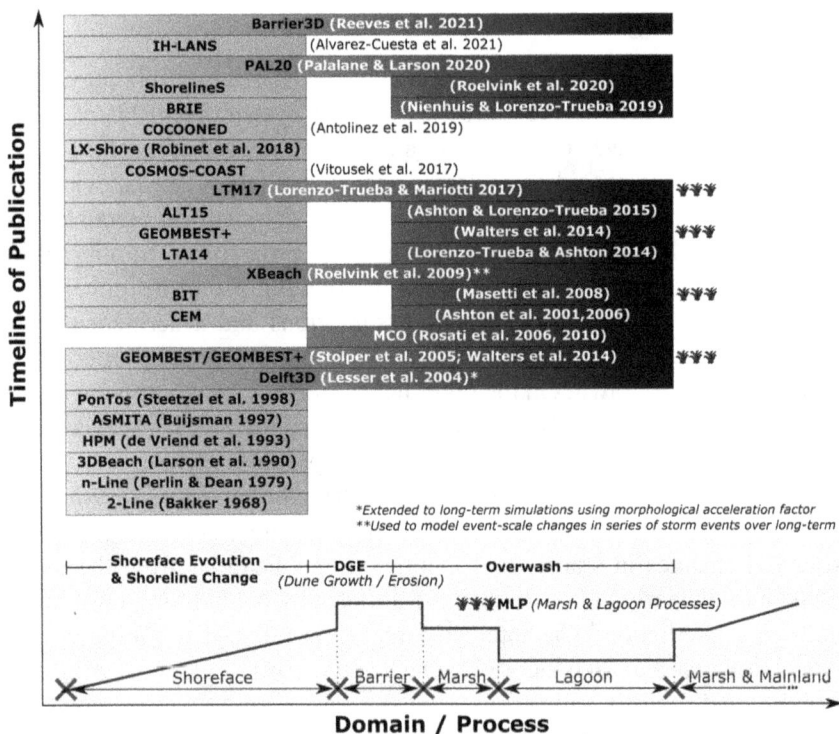

Fig. 6.   Long-term morphodynamic models with a coupled approach. Models are shown according to their publication chronology and are aligned with their respective processes, which range from shoreface erosion or shoreline change, to dune growth/erosion, to overwash.

Table 4. Long-term morphodynamic models.

Year	Model Name [Ref.]	Modeled Phenomena/Processes†							
		SF	SL	TR	ID	DC	SB	OW	ML
1956	PEL56 (Pelnard-Considere, 1956)		X						
1962	Bruun Rule (Bruun, 1962)			X					
1968	2-Line (Bakker, 1968)	X	X						
1979	n-Line (Perlin and Dean, 1979)	X	X						
1983	Gen. Bruun Rule (Dean and Maurmeyer, 1983)			X					
1985	EVR85 (Everts, 1985)			X					
1989	GENESIS (Hanson and Kraus, 1989)		X						
1990	3DBeach (Larson et al., 1990)	X	X						
1992	STM (Cowell et al., 1992)			X					
1993	HPM (de Vriend et al., 1993)	X	X						
1995	ADM (Niedoroda et al., 1995)	X							
1997	ASMITA (Buijsman, 1997)	X	X		X				
1998	PonTos (Steetzel et al., 1998)	X	X						
2001	CEM (Ashton et al., 2001; Ashton and Murray, 2006)		X					X	
2002	Cascade (Larson, 2002)		X		X				
2002	BARSIM (Storms et al., 2002)			X				X	
2005	GEOMBEST (Stolper et al., 2005)	X		X				X	X
2006	MCO (Rosati et al., 2006, 2010)			X		X	X	X	
2008	BIT (Masetti et al., 2008)	X		X				X	X

(*Continued*)

Table 4.   (*Continued*)

Year	Model Name [Ref.]	Modeled Phenomena/Processes[†]							
		SF	SL	TR	ID	DC	SB	OW	ML
2009	YAT09 (Yates et al., 2009)	X							
2012	GenCade (Frey et al., 2012)		X		X				
2013	ShoreFor (Davidson et al., 2013)	X							
2013	Mod. Bruun Rule (Rosati et al., 2013)			X				X	
2014	LTA14 (Lorenzo-Trueba and Ashton, 2014)	X		X				X	
2014	GEOMBEST+ (Walters et al., 2014)	X		X				X	X
2015	ALT15 (Ashton and Lorenzo-Trueba, 2015)	X	X	X				X	
2016	D&H16 (Dean and Houston, 2016)			X	X				
2017	LTM17 (Lorenzo-Trueba and Mariotti, 2017)	X		X				X	X
2017	CoSMoS-COAST (Vitousek et al., 2017)	X	X	X					
2018	LX-Shore (Robinet et al., 2019)	X	X						
2019	COCOONED (Antolinez et al., 2019)	X	X	X		X			
2019	BRIE (Nienhuis and Lorenzo-Trueba et al., 2019)	X	X	X	X			X	
2020	ShorelineS (Roelvink et al., 2020)		X					X	
2020	PAL20 (Palalane and Larson, 2020)	X	X	X	X	X		X	
2021	UNIBEST-CL+ (Deltares, 2021)	X	X						
2021	ShoreTrans (McCarroll et al., 2021)			X		X			
2021	IH-LANS (Alvarez-Cuesta, 2021)	X	X						
2021	Barrier3D (Reeves, 2021)	X		X		X		X	

Notes: [†]SF: Shoreface Change; SL: Shoreline Change; TR: Transgression; ID: Inlet Dynamics; DC: Dune Changes; SB: Subsidence; OW: Overwash; ML: Marsh and Lagoon Processes.

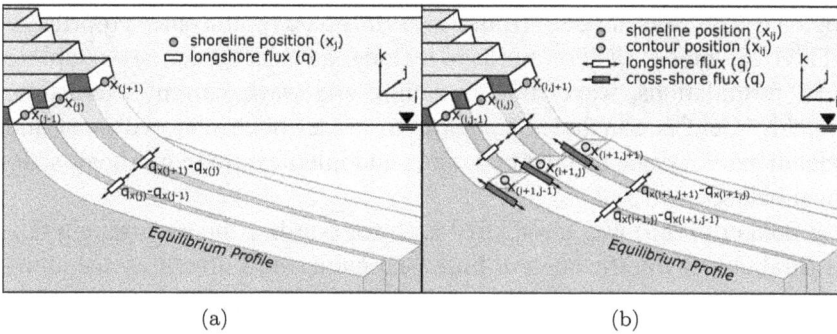

Fig. 7. One-line and two-line model schematics: (a) One-line approach that predicts shoreline changes based on LST gradients $(qx_{(j+1)} - qx_{(j)})$. (b) Two-line approach that predicts change at the shoreline and an offshore contour, considering LST gradients in each zone and rule-based XST.

Source: Figure modified from the work of Perlin and Dean (1979).

modeling shoreline evolution stemmed from one-line theory, published by Pelnard-Considère (Latteux, 1995). Models derived from this theory, commonly called "one-line models," assume a constant equilibrium profile and calculate position changes in a single contour line — the shoreline — over time considering only the gradients in the LST rate (Fig. 7(a)).

Larson *et al.* (1987) published a review of one-line modeling theory and analytical solutions that had been developed for various coast-specific and structure-specific situations. Two years later, Hanson and Kraus, (1989) presented the one-line model GENESIS, which would become one of the most widely used one-line models for predicting shoreline evolution in practice, though not without criticism, (e.g., Young *et al.*, 1995; Houston, 1996). One-line models are still being developed and used today, likely due to their simplicity, intuitiveness, and ease of calculation. The Coastal Evolution Model (CEM) of Ashton *et al.* (2001) is a one-line model that predicts shoreline response due to high-angle waves, assuming a constant linear shoreface out to an estimated closure depth. From numerical experiments, they found that high-angle waves cause small shoreline perturbations to grow into larger formations, such as cuspate and spits. Additionally, they found that shoreline protrusions can shelter downdrift features from the high-angle waves, affecting the evolution of such features. Thomas and Frey (2013) and Kim *et al.* (2020) reviewed other common one-line models including UNIBEST-CL+ (2021) GenCade, (Frey *et al.*, 2012) which is a combination of GENESIS and the regional Cascade model, (Larson et al., 2002) and the proprietary LITPACK model. These models include advances such as coupling XST formulations, wave transformation, and wave–current interaction. Notably, GenCade includes advances to model tidal inlet evolution and inlet dynamics, such as inlet bypassing and inlet feature (e.g., shoal) sediment balance.

Bakker (1986) was unsatisfied with the one-line theory's assumption of parallel bathymetric contour lines near engineered structures due to the apparent discontinuity it produced. In 1968, Bakker published a two-line model whereby XST could be approximated between two profile zones based on the profile's deviation from an equilibrium state (Fig. 7(b)). Perlin and Dean (1979) were the first to suggest expanding Bakker's two-line approach to multiple lines and followed up with publication of their n-line model six years later, which was named for its ability to handle a user-defined 'n' number of contour lines (Perlin and Dean, 1985). Although limited in their ability to produce non-monotonically decreasing profiles, these models were the first to add elements of cross-shore change to one-line models, paving the way for later n-line models that would attempt to integrate both XST and LST, (e.g., Steetzel et al., 1998).

Buijsman (1997) published the ASMITA model, which simulated interaction between the adjacent shoreline and tidal inlets. The model consisted of five nodes that represented the tidal channel, ebb shoal, flood shoal, and the adjacent shorelines. Sediment flux between these nodes was calculated based on equilibrium formulations of each feature. A similar approach was incorporated into the regional barrier island model called Cascade, presented by Larson et al. (2002). While ASMITA focused on modeling the channel evolution, Cascade focused on modeling the regional shoreline position over long time scales but accounted for the dynamic inlet features in the form of sediment source and sink terms. Larson et al. (2002) applied Cascade to a regional stretch of a U.S. East Coast barrier island and found the model was able to satisfactorily predict the shoreline position updrift and downdrift of two inlets.

3.2.2. Modeling shoreface evolution

Although long-term modeling of barrier transgression was well underway by the 1980s, most models assumed a constant profile shape. It wasn't until the mid-1990s that shoreface evolution began to be modeled, with the publication of the Hinged Panel Model (HPM) (Vriend, 1993) and the Advection– Diffusion Model (ADM) (Niedoroda, 1995).

A conceptualized model of the shoreface profile by de Vriend *et al.* (1993) discretized the shoreface into three sections: (1) the upper shoreface, (2) the lower shoreface, and (3) the middle shoreface, which acted as a transition zone between the upper and lower zones. On the lower shoreface, profile movement was assumed to be negligible compared to the scales of interest, while the upper shoreface was assumed to be highly active out to the depth of closure (i.e., the transition point to the middle shoreface). The sections were considered to be rigid panels, which rotated about hinge points at the panel intersections based on the net sediment transport into or out of the panel zone. This led Cowell *et al.* (2003) to refer to this model as the Hinged-Panel Model (HPM). Stive *et al.* (1995) published a full treatment on HPM, which used Bowen's energetics formulation for XST between the shoreface sections. They found that HPM produced reasonable hindcast simulations and that the effect of substrate slope on profile evolution was only relevant at geologic timescales.

Niedoroda *et al.* (1995) published a similar model, the main difference being the continuous formulation of XST as compared to the paneled formulation of Stive and de Vriend. The continuous formulation is depth-dependent and breaks down the transport into a bed load (i.e., advective) term and a suspended load (i.e., diffusive) term; thus, it was called the ADM model by Cowell *et al.* (2003). Although Stive *et al.* (1995) and Niedoroda *et al.* (1995) do not apply their models to barrier coasts specifically, their work signifies advancement in cross-shore shoreface modeling and the increased importance of including cross-shore processes in long-term models.

Another class of models that simulate shoreface evolution are equilibrium shoreline models, which have become increasingly popular for simulating event-based to interannual change. These models combine equilibrium-based formulations of shoreface evolution with shoreline change models (typically one-line models). The two most popular models include the model of Yates *et al.* (2009) and the Shoreline Forecast (ShoreFor) model of Davidson *et al.* (2013). Both models demonstrate that beaches often respond directly to wave forcing (e.g., as quantified by wave energy or dimensionless fall velocity); however, the equilibrium response time scale (which is often longer than a single storm event) plays an exceedingly important role in the morphological evolution. Further, the extensive observations and developed model of Yates *et al.* (2009)

show that beaches become increasingly resistant to erosion while in an eroded state.

3.2.3. *Modeling barrier transgression*

Models of shoreline change and shoreface evolution often produce a landward or seaward shift in the shoreline and/or profile based on gradients in the sediment transport rates. However, these models are not able to account for barrier transgression as an observed phenomenon. Thus, numerous models were developed to simulate long-term transgression based on cross-shore processes (e.g., overwash, breaching, and inlet dynamics) and long-term forcing conditions (e.g., SLR).

Translation models
Bruun (1962) introduced what is perhaps the most popular hypothesis about cross-shore transgression, which states that an equilibrium beach profile translates upward and landward with SLR while conserving sediment volume. Years later, this became known as the "Bruun Rule (1983)." Since the profile is "translated," these types of models are often called "translation models" in the literature, and many of them have been developed since the publication of the Bruun rule.

The Bruun rule (1983) predicts profile recession distance based on the amount of SLR and the average beach slope while conserving sediment. In subsequent examination of his theory, Bruun (1983) revisited the assumptions behind the model development and cautioned modelers who might attempt to apply the Bruun rule in coupled alongshore models and progradational scenarios. Upon further review of initial publications by Bruun (1983) and Schwartz (1967) several researchers have offered criticism of the way that the Bruun rule (and the underlying equilibrium profile concept) is used in current models, (e.g., Pilkey *et al.*, 1993; Thieler *et al.*, 2000). Conceding that some of the criticisms of Pilkey *et al.* (1993) were valid, Dubois (1993) stated that such models can still be useful in formulating research questions and site-specific equilibrium-based models. A more recent study by Wolinsky and Murray (2009) highlighted additional limitations of the Bruun rule as applied to long-term simulations on the order of millennia.

Rosati *et al.* (2013) offered a review of field studies that attempted to validate the Bruun rule (or modified forms of it). More recently, the Bruun rule has been used to model both barred and bermed beach profiles in a laboratory setting, (e.g., Atkinson *et al.*, 2018). D'anna *et al.* (2021) recently presented a reinterpretation of the Bruun rule that explicitly partitions shoreline recession into passive flooding of the beach profile and wave-driven reshaping components. Similarly, Troy *et al.* (2016) assessed long-term profile submergence versus Bruunian recession of beaches on the Great Lakes, a model environment to observe the effects of significant water level variability, which serves as a proxy for future SLR.

The Bruun rule has also been expanded since its initial publication. Dean and Maurmeyer (1983) presented the generalized Bruun rule, which expanded the original model to include the recession of barrier coasts specifically, and noted that greater recession rates were predicted due to the additional sand volume being deposited on the subaerial island and in the lagoon. The Bruun rule was also expanded to include source and sink terms in the models of Everts (1985, 1987). Everts proposed that historical rates of SLR and shoreface retreat are preserved in the slope of the seaward profile, assuming that the profile is not significantly reworked by LST or tectonic deformation processes. Everts compared present and past ratios of SLR to shoreface retreat for five U.S. East Coast barrier islands and found that some barriers are in a narrowing state. Everts proposed that these barriers would continue to narrow until a critical width is reached, at which point landward migration of the island would begin. This theory employed the previously mentioned critical length concept, which was first proposed in (Leatherman, 1983) and has since been utilized in other models, (e.g., Lorenzo-Trueba and Ashton, 2014). Further modifications of the Bruun rule were published by Rosati *et al.* (2013) who included an additional term representing XST in the landward direction by overwash and/or aeolian processes, and Dean and Houston (2016) who added a LST term and sediment source/sink terms to Rosati's 2013 formulation.

Cowell *et al.* (1992) developed the Shoreface Translation Model (STM), which allowed modelers to keep track of changes in stratigraphy and was later used in conjunction with field observations to perform hindcasting simulations (Cowell *et al.*, 1995). The STM was later expanded using a probabilistic framework to produce distributions of results that could be statistically evaluated in risk management frameworks (Cowell *et al.*, 2006).

Most recently, McCarroll *et al.* (2021) published the ShoreTrans model, which follows a similar profile translation methodology with a couple of distinctions and additions. First, the model uses measured profiles instead of parametric representations. Second, in addition to the profile translation, ShoreTrans has been modified to incorporate dune erosion and accretion, sediment flux between the upper (active) and lower (inactive) shoreface, as well as source and sink terms that can modify the sediment supply.

Other transgression models
More recent transgression models can't simply be described as 'translation' models since they also simulate profile changes. For example, Storms *et al.* (2002) published an evolution model called BARSIM, which was intended to preserve the simulation's erosion and depositional time history for comparison to observed shoreface stratigraphy. They describe BARSIM as a 'process–response' model in which erosional and depositional mechanisms were modeled separately. Storms *et al.* (2002) conducted multiple numerical experiments and found that their model successfully captured several general observations: (1) increased grain sizes led to steeper shoreface slopes, (2) higher sediment supply values decreased retrogradation and increased the likelihood of aggradation or progradation, (3) higher SLR rates increased the likelihood of barrier overstepping, and (4) lower substrate slopes allowed for greater landward rates of migration.

Stolper *et al.* (2005) published the GEOMBEST model, which allows for depth-dependent shoreface adjustment toward a theoretical equilibrium profile, thus allowing the shoreface to temporarily exist in disequilibrium. GEOMBEST is also able to simulate heterogeneous stratigraphic units that can differ in erodability. Using the conceptual model of Cowell *et al.* (2003) GEOMBEST divides each simulated coastal tract into three cross-shore zones (i.e., shoreface, backbarrier, and estuary). Stolper *et al.* (2005) used this model to estimate possible stratigraphic histories in both steep and gentle sloping environments, showing that quantitative estimates may be useful where historical data may be lost or otherwise unavailable. They also showed that substrate slope plays an important role when non-erodable outcrops are present. Specifically, they found that steep slopes lead to the narrowing of the estuary and barrier drowning unless there is an external increase in sediment supply.

Based on sensitivity analyses with GEOMBEST, Moore *et al.* (2007) found that increasing the SLR rate and decreasing sediment supply led to increased barrier migration. Moore *et al.* (2010) also studied the

Holocene evolution of U.S. East Coast barrier islands and found that the most vulnerable islands were large with less erodable substrates and gentle slopes. Brenner *et al.* (2015) confirmed these findings and also found that positive and negative feedbacks occur based on the slope of the substrate and island trajectory, and the composition of the substrate and backbarrier deposits; the negative feedback adjusts the island trajectory to the substrate slope while the positive feedback leads to barrier width adjustments.

In studying the effects of compaction on barrier island migration, Rosati *et al.* (2006) developed the Migration, Consolidation, and Overwash (MCO) model to predict the response of barrier systems to a series of storm events. The MCO model used the Convolution Method of Kriebel and Dean (1993) to predict responses when there was no over-wash and the numerical method of Donnelly *et al.* (2005) to estimate overwash volumes when water levels exceeded the berm height. Rosati *et al.* (2006) found that when consolidation was considered, there were considerable increases in migration distance and reduction of dune elevations. They found that increases in surge heights and deep-water wave heights also led to significant increases in migration and reduction of dune elevations. Rosati *et al.* (2010) updated the 2006 model to include the overwash formulations by Donnelly *et al.* (2009) and found that barriers on top of compressible substrates migrated much faster than barriers on non-compressible substrates, assuming a sufficient sand supply. They also found lower dune elevations and island volume loss to be more prevalent when compressible substrates were present, the thickness of which was found to be nonlinearly related to consolidation rates.

Masetti *et al.* (2008) developed the Barrier Island Translation (BIT) model with separate sediment transport formulations for shoreface evolution, inner shelf reworking, overwash, and backbarrier infilling. They found barrier migration to undergo significant increases and decreases in migration rate according to the substrate slope and sediment availability. Additionally, they found that offshore subaqueous bodies of sediment were most likely due to barrier migration over a non-uniform surface rather than drowning of previous barrier islands.

Lorenzo-Trueba and Ashton (2014) developed a barrier island evolution model (hereafter the "LTA14" model) to evaluate the long-term behavior of the system. The model tracked transect boundary changes in the cross-shore direction based on sediment flux calculations. They found that barriers evolved following one of the four behaviors: height drowning, width drowning, constant transgression (or dynamic equilibrium), and periodic transgression. Most recently, Reeves *et al.* (2021) expanded the LTA14 model domain to consider dune and subaerial island processes in a model called Barrier3D. The Barrier3D model used the LTA14 equations to simulate shoreline and nearshore profile change and included additional formulations for dune growth during non-stormy periods, dune reduction by overwash, alongshore dune elevation changes, and sediment transport by overwash and backbarrier overland flow. Barrier3D also used probability distributions to simulate synthetic storm events and barrier recovery between storms (Reeves *et al.*, 2021).

3.2.4. *Multifaceted evolution models*

Whereas most of the previously discussed long-term models were developed to simulate one primary phenomenon (e.g., shoreline change, shoreface evolution, and barrier transgression), other recent models have been developed with the intent to simulate multiple long-term phenomena. We discuss four categories of these multifaceted evolution models: (1) coupled barrier–backbarrier models, (2) models that combine shoreline change and transgression, (3) models that combine shoreline change and shoreface evolution (i.e., equilibrium shoreline models), and (4) extended event-scale models.

Coupled barrier–backbarrier models
In the last decade, barrier island evolution models have been coupled with backbarrier models to evaluate interactions or feedback between the systems. Walters *et al.* (2014) published GEOMBEST+, which coupled GEOMBEST with a backbarrier model of Mariotti and Fagherazzi (2010). Using this model, they found that overwash played an important role in that it provided a narrow platform for backbarrier marsh growth, which in turn reduced island migration rates by decreasing accommodation space for sediment deposition. Lorenzo-Trueba and Mariotti (2017) also developed a coupled model that combined the backbarrier marsh model of Mariotti and Carr (2014) and Lorenzo-Trueba and Ashton (2014). They found that including processes such as the import/export of fine sediment to the barrier environment significantly impacted the accommodation space for overwashed sediment, which ultimately led to either a sustained island that migrated or one that drowned.

Models that couple shoreline change and transgression

Noting that most of the previous modeling efforts focused on either shoreline change or transgression, models are increasingly being developed to include both components. In 2006, the CEM model was updated to include a function for barrier overwash (Ashton and Murray, 2006) and was later coupled with the LTA14 cross-shore barrier model (Ashton and Lorenzo-Trueba, 2015). The authors found that when alongshore coupling was less significant, large alongshore variations persisted longer in the simulation; thus, alongshore coupling was found to act as a dampener on barrier transgression (Ashton and Lorenzo-Trueba, 2015).

Nienhuis and Lorenzo-Trueba (2019) published the BarrieR Inlet Environment (BRIE) model, which modified and extended the combined model of Ashton and Lorenzo-Trueba (2015) to include inlet dynamics. The model simulated inlet formation (i.e., breaching) and cross-sectional area changes, including alongshore sediment volume balancing between updrift and down-drift sides of the inlet. BRIE also included a stratigraphic model that keeps track of how sediment types (i.e., lagoonal, washover deposits, and flood tidal shoals) are reworked over time Nienhuis and Lorenzo-Trueba (2019).

Other models include that of Palalane and Larson (2020) ShorelineS, (Roelvink *et al.*, 2020) and IH-LANS (Alvarez-Cuesta, 2021). The Cascade model, which simulates shoreline changes for a region of barrier islands, was updated by Palalane and Larson (2020) to include XST components from Larson *et al.* (2016) which included overwash, beach and dune erosion, transport between the beach and offshore bar, and aeolian transport. The ShorelineS model, developed by Roelvink *et al.* (2020) models shoreline change, overwash, and includes the ability to split and merge barrier islands or spits. It is also planned for ShorelineS to be coupled with XBeach or Delft3D to simulate island and inlet migration in future work Roelvink *et al.* (2020). Alvarez-Cuesta *et al.* (2021) developed the IH-LANS model which combines LST (using a modified version of CERC based on Hallermeier, 1980) and XST (following Toimil *et al.*, 2017), while also including specific formulations for engineering structures, such as groins, seawalls, and breakwaters.

Models that couple shoreline change and shoreface evolution

Although not limited to barrier island modeling, many long-term models now couple shoreline change and shoreface evolution models. One of the earliest examples of this approach was the 3DBeach model, published by

Larson *et al.* (1990) which was a combination of SBEACH and GENESIS and was capable of simulating dynamic profile features, such as offshore bars.

Recently developed models incorporate equilibrium shoreline models as one aspect of their predictive capabilities. These models include CoSMoS-COAST, (Vitousek *et al.*, 2017) LX-Shore, (Robinet *et al.*, 2017) and COCOONED (Antolínez *et al.*, 2019). CoSMoS-COAST combines the one-line model of Vitousek and Barnard (2015) the equilibrium model of Yates *et al.* (2009) a translation component similar to Bruun (1962) and a long-term residual shoreline trend following Long and Plant (2012). LX-Shore combines the wave model SWAN with LST, (e.g., Kamphuis, 1991), and XST, (e.g., Davidson *et al.*, 2013), formulations in a 2D horizontal grid, similar to the CEM model setup (Robinet *et al.*, 2018). Lastly, the COCOONED model (Antolínez *et al.*, 2019) couples a one-line approach similar to Vitousek and Barnard (2015) a cross-shore equilibrium model similar to Miller and Dean (2004) and the analytical dune erosion method of Kriebel and Dean (1993).

Notably, data assimilation techniques have been tried with many of these equilibrium shoreline models. Long and Plant (2012) were one of the first to use data assimilation for shoreline evolution predictions. They combined a modified version of the Yates *et al.* (2009) model, which predicts long-term and short-term trends of shoreline position, with a joint extended Kalman Filter (eKF) assimilation approach that updates the model predictions based on shoreline position observations. Other models that have used Kalman filtering include CoSMos-COAST, (Vitousek *et al.*, 2017) ShoreFor, (Ibaceta *et al.*, 2020) and IH-LANS (Alvarez-Cuesta *et al.*, 2021).

Extended event-scale models

Another common modeling approach that combines XST and LST is the extension of multifaceted event-scale models for use in long-term simulations. Due to computational constraints, event-scale models have primarily been used to simulate short-term changes. However, recently they have also been employed and extended to predict long-term changes where computational burden is reduced through hydrodynamic averaging or lengthening the morphological time step.

Vemulakonda *et al.* (1988) were among the first to utilize this approach with the Coastal Inlet Processes (CIP) model, which was originally developed to predict tidal inlet shoaling for ingress and egress of

U.S. submarines. Wave and circulation models were coupled together with a sediment transport model, the latter of which required a user-defined time step that effectively extended the hydrodynamic conditions. Comparing model results to a year's worth of navigation channel survey data, the model was shown to satisfactorily predict sediment transport rates (Vemulakonda *et al.*, 1998).

A more recent and common approach is that of Lesser *et al.* (2004) who applied a morphological acceleration factor (*morfac*) within Delft3D to effectively lengthen the sediment transport time step for long-term simulations. Lesser *et al.* (2004) showed that using *morfac* in simplified cases did not cause the results to significantly deviate from the full solution. This approach was extended by Roelvink (2006) who proposed running multiple accelerated simulations in parallel for different tidal phases and using a weighted average of morphological change to update the bathymetry for the next time step.

Event-scale models are also used to model storm sequences and recovery periods between storms. Ranasinghe *et al.* (2012) developed the Probabilistic Coastline Recession (PCR) model, which generates 100-year sequences of storm events and employs the event-scale swash impact model of Larson *et al.* (2004) (LEH04) to predict dune recession. The model also considered SLR projections and used a constant, empirically derived rate of dune recovery between storm events (Ranasinghe *et al.*, 2012). Long *et al.* (2020) developed a modeling framework for Breton Island, Louisiana, to assess restoration design alternatives that used XBeach to model the island's response to successive storm events over a 15-year time period. Shoreface and bay-side erosion between storm events were not modeled explicitly but were accounted for through manual manipulation of the pre-storm profiles (Long *et al.*, 2020).

3.3. *Summary of advancements and limitations*

The literature indicates that over the last 50 years, significant advancements have been made in long-term morphodynamic modeling of barrier systems. Again, model development has followed a rather natural progression — from the simplified to the complex. The intuition behind some of the earliest models (e.g., one-line and translation models) laid a foundation on which subsequent model development has been steadily built. More complex formulations have been developed to predict shoreface shape changes rather than assuming a constant equilibrium profile. Additional processes have been added (e.g., overwash representations and changes in sediment supply) to more closely capture the underlying mechanics of barrier transgression. Models are also increasingly being developed to incorporate other sub-systems (e.g., the backbarrier marsh–lagoon system) that impact the long-term morphodynamics.

Yet, there are still many limitations to be addressed, including (but not limited to) model validation, uncertainty characterization, and the incorporation of relevant processes and important factors. Although there is a wealth of satellite imagery available to coastal researchers, this dataset is limited both in the information it contains (i.e., primarily shoreline and marsh positions) and its temporal coverage for long-term model calibration and validation. This lack of long-term quantitative data is one likely reason why many long-term models have not been thoroughly validated. Other long-term models that were originally created to explore barrier island morphodynamics and develop new hypotheses — what Murray (2003) calls "exploratory models" — have largely remained as such and have not yet shifted toward the prediction of real systems. Additionally, although testing model sensitivity is common practice, most models are not developed to explicitly consider input parameter uncertainty. Models typically receive averaged or representative input values and produce a single-value output rather than a statistical range of predictions. Another limitation, similar to event-scale modeling, is that most previous efforts have focused on evolution of the natural barrier system and have neglected anthropogenic impacts. Other relevant processes such as barrier subsidence, aeolian transport, backbarrier marsh growth/erosion, and factors that impact erosion and deposition such as vegetation type and density have mostly been excluded from long-term models with only a few exceptions.

One modeling challenge that has persisted over time is the extrapolation of small-scale sediment transport predictions to large-scale coastal behavior (LSCB) — a link which is certainly intuitive. However, the problem of uncertainty or error propagation, where uncertainty or error at the small scale compounds over time resulting in imprecise or inaccurate predictions, has stifled this type of long-term modeling. De Vriend (1991) indicates the extraordinary challenge of this unsolved problem saying, "...it must even be doubted whether models formulated at a small scale will ever be able to describe LSCB," and reverently quips that "we may need another Ludwig Prandtl" before we have a good answer.

4. Research Gaps and Needs

Based on the advancements that have been made toward our *Grand Challenge* and the limitations that persist in our modeling efforts, we have identified critical gaps and future research needs that might be addressed moving forward. The gaps and needs highlighted in the following are those we believe are most critical for making progress toward the *Grand Challenge*. We acknowledge, however, that other gaps and needs exist. The research gaps and needs may be generally categorized as follows: (1) observations, data availability, and accessibility, (2) scientific understanding of relevant processes, and (3) modeling framework and approach. These categories are expounded in the following.

4.1. *Observations, data availability, and accessibility*

One of the major limitations of our current modeling efforts is the availability of data. While technological advancements during the 20th century increased our ability to collect good data, the timing of these advancements means the quantity of long-term data for validation is sparse. On the other hand, event-scale data are not limited by time but by the complexities and dangers associated with collecting perishable data before, during, and immediately following storm events. However, to improve our scientific understanding of the relevant processes and associated modeling efforts, we must overcome these data limitations so that we can ground truth in our theories and formulations in observations. Herein, we discuss a few high-level issues regarding data acquisition and accessibility while assuming that some methodological advancements for data collection and analysis will be required to further our understanding of the relevant processes discussed in the following section.

Long-term observations of coastal morphodynamics generally exist only at a limited number of well-monitored sites (e.g., Duck, NC (Larson and Kraus, 1994); Torrey Pines, CA (Ludka *et al.*, 2019); Ocean Beach, CA (Barnard *et al.*, 2012); Fire Island, NY (Lentz and Hapke, 2011); Narrabeen-collaroy, Australia (Turner *et al.*, 2016); Truc Vert, France (Castelle *et al.*, 2020); Hasaki, Japan (Banno *et al.*, 2020); and South Holland, Netherlands (de Schipper *et al.*, 2016)), which are maintained by various government agencies and academic institutions. It is vital that these long-term monitoring efforts continue while new avenues of data at higher spatiotemporal resolutions are sought. As such, we must be diligent in making the most of available datasets, develop new ones, and make them broadly accessible. We must develop and promote centralized, open-access databases (e.g., the Community Surface Dynamics Modeling System — CSDMS) that contain both open-access models and collected data (e.g., the use of public archival in the National Science Foundation's DesignSafe (Rathje *et al.*, 2017) or post-event field data (Berman *et al.*, 2020). Increasing the amount and quality of available data would also be useful for blind model comparisons, data assimilation, and machine learning applications.

One way to push toward increased dataset availability is to continue to capitalize on technologies that exist and are readily available. A perfect example of this is remote sensing data, such as publicly available satellite imagery, (e.g., Luijendijk *et al.*, 2018; Vos *et al.*, 2019; Turner *et al.*, 2021). We also expect that publicly accessible LiDaR datasets will become more widely available with continued advancements in drone technology (Shaw *et al.*, 2019). It might also require us to creatively enlist the public's help in data collection, such as using public photos and photogrammetry, (e.g., Harley *et al.*, 2019). A second way to advance this initiative is by developing new data collection methods or technologies. Due to the perishable nature of pre- and post-storm data and the uncertainties surrounding the timing and location of storm events, morphological data before, during, and after storm events is difficult to obtain. Certain efforts are underway to help coordinate, collect, and make available this perishable data, including the National Science Foundation's NHERI RAPID Facility (Berman *et al.*, 2020; Wartman *et al.*, 2020) and Nearshore Extreme Events Reconnaissance program (Raubenheimer, 2020).

4.2. *Scientific understanding of relevant processes*

Epistemic uncertainty and the exclusion of relevant factors are two important previously mentioned limitations. The epistemological issues discussed herein include both hydrodynamics and sediment transport, and the relevant factors discussed include vegetation dynamics and anthropogenic impacts.

Despite hydrodynamic simulation advancements, increased complexity in sediment transport formulations has not always translated to increased accuracy. Quoting from a study by Davies *et al.* (2002) in which multiple transport formulations were compared, Bosboom and

Stive (2021) noted that most sediment transport predictions are only accurate within an order of magnitude and that empirical calibration of these model formulations is still necessary in many cases. They also remarked that the simpler formulations are still often the best available ones. This indicates an obvious shortcoming in our ability to reproduce realistic hydrodynamic forcing conditions and to model the relationship between forcing and sediment transport. Aagaard and Hughes (2013) highlighted some of the latter shortcomings, stating that there is room for improvement in our quantitative understanding of bed load and suspended load transport, as well as our knowledge of which parameters (other than bed shear stress) can lead to better transport rate predictions. Notably, while such improvements would certainly lead to advancements in event-scale modeling efforts, the initial impact on long-term models would be minimal.

One of the greatest advancements in event-scale morphodynamic modeling in recent years was the inclusion of infragravity waves in hydrodynamic calculations (Sherwood, 2020). While we still do not fully understand the mechanics of how these waves impact nearshore sediment transport (Aagaard and Kroon, 2017), we now recognize their importance in predicting event-scale morphodynamic response. Other factors such as the nonlinearity of incident waves, the interaction of incident and infragravity waves, and swash zone dynamics, including turbulence and boundary layer flows, may also prove to be key missing components in coupled hydrodynamics–sediment transport formulations that have a significant impact on event-scale morphodynamics. While these factors may be key missing components, the small scales needed to resolve some of these hydrodynamic and sediment transport processes would require computational resources that make such modeling practically infeasible at present. Continued computational advancements may help alleviate such limitations.

In studying and developing formulations for event-scale processes such as overwash and breaching, it is important to consider all of the contributions to total inundation height, including tides, storm surges, and waves. The exclusion of one or more of these contributions can alter the total inundation height and corresponding morphological response. Furthermore, special consideration should be given to the timing of these contributions, as recent work has shown that time differences between the bay peak surge and ocean peak surge can lead to bay-side breaching, (e.g., Shin, 1996; McCall *et al.*, 2010; Smallegan and Irish, 2010).

Since data for event-scale morphodynamic response are sparse, future work should capitalize on previously published studies or available data from historical events, (e.g., van Ormondt *et al.*, 2020), which may yield additional insights into the nature of overwash and breaching. Moreover, since overwash and breach observations are difficult to obtain in the field, physical modeling that leverages advancements in data collection methods and instrumentation may also help us better understand and quantify these processes. Although these physical modeling studies would require careful consideration of potential scaling issues, we believe that valuable insights into the overwash and breaching processes remain to be gained from this method of study.

Another factor that may be prioritized for future studies is coastal vegetation. Currently, we have a general understanding of how vegetation impacts barrier morphodynamics (e.g., dune stabilization, subaerial accretion, and increased flow roughness) and vice versa, (e.g., van der Lugt *et al.*, 2019); however, our quantitative understanding, and field-verification of that understanding, is further behind. Moving forward, beneficial research efforts would include the quantification of vegetation impact for parameters such as vegetation type, location, density, and hydrodynamic conditions for implementation in event-scale and long-term models. Recent studies, (e.g., Ayat and Kobayashi, 2015; Zinnert *et al.*, 2019), indicated that this research is underway, and recent modeling studies, (e.g., Passeri *et al.*, 2018; van der Lugt *et al.*, 2019), exemplify the initial stages of incorporating this information into event-scale morphodynamic analysis. Furthermore, with the U.S. Army Corps of Engineers' recent release of international guidelines on the design and implementation of Natural and Nature-Based Features (NNBF) (Bridges *et al.*, 2021), we expect future studies to quantify the performance of NNBF in various coastal environments.

Many coastal barriers are no longer representative of a natural environment as they are either developed or impacted by development and engineering structures on neighboring shorelines. Although many early studies and models sought to quantify the impact of engineering structures on littoral transport (e.g., one-line modeling of shoreline changes near groins), relatively few studies have quantitatively addressed the morphological impact of human development and other large-scale coastal restoration practices. Additionally, we would benefit from better understanding of how the coastal management process works holistically, including how policies are developed, how individual restoration decisions are made, and how studies which quantify anthropogenic impacts influence the management process, considering cultural, political, and socioeconomic differences across localities. This type of analysis has largely been absent in the barrier morphodynamics literature, with the exception of a few observational studies on the feedback between coastal protection and real estate values, (e.g., Keeler *et al.*, 2018), and modeling studies that consider the coupling of barrier morphodynamics with the incentives of developers and owners, (e.g., McNamara and Werner, 2008), and individuals in the coastal real estate market, (e.g., McNamara and Keeler, 2013). Moving forward, beneficial research topics would include understanding the quantitative morphodynamic response between developed and natural barrier systems, (e.g., Rogers *et al.*, 2015), and the incentives, behavior, and impacts of human agents in what is appropriately called a "coupled human–landscape" or "coupled natural–human" system (McNamara and Lazarus, 2018; NASEM, 2018).

4.3. *Modeling framework and approach*

There are several ways in which our modeling frameworks and overall approach may continue to improve in order to further research and achieve higher-confidence predictions. First, since modeling is inherently tied to the scientific understanding of the processes being studied, advancements in how those processes are understood must be regularly incorporated into the improvement of existing models and the development of new models. As research has naturally become more focused and specialized, many recent studies have been published related to specific components of barrier island morphodynamics (e.g., sediment transport between the inner shelf and active profile, beach–dune interactions, and backbarrier marsh dynamics). Therefore, it is critically important that holistic models of barrier morphodynamics incorporate the theory and formulations of more focused models.

Second, although some of the recently published long-term morphodynamic models included sensitivity analyses for various parameters, model results are still largely presented as a single simulation output. Modeling efforts would benefit by increasingly employing ensemble approaches (e.g., Monte Carlo techniques) that consider input parameter uncertainty. Rather than producing a single output, a probabilistic range of results would be produced that can help characterize uncertainty in the model predictions (Vitousek *et al.*, 2021). Such an approach lends itself to identifying not only expected values but also extreme scenarios and the input parameter combinations that cause them. Additionally, with the large number of models that have been developed, modelers may consider a multiple-model ensemble approach to evaluate the range of predictions across various models, as has been done with model comparison studies, (e.g., Montaño et al., 2020). Such an approach would emulate the current practice for forecasting hurricanes and would also naturally facilitate model comparisons and identification of robust and accurate models.

Third, as we focus on expanding data accessibility and collection capabilities, we must be diligent to incorporate the available data. In addition to model validation, data may be used to train and/or reduce errors in model predictions using machine learning and data assimilation methods, respectively. There are many ways in which machine learning may be employed in morphodynamic modeling to improve predictions and fine-tune model parameters for a specific site (Goldstein et al., 2019). Machine learning may also be employed to reduce the computational burden. As models include relevant processes at smaller scales, the computational burden will naturally increase; however, machine learning techniques can serve to abstract those computationally expensive processes, effectively substituting a recognized or learned pattern for a more complex algorithm. One drawback to these powerful data-driven approaches is that it is possible to "over-train" a model with limited data, which effectively reduces its predictive capability for conditions that have yet to be observed. Despite the benefits and drawbacks of these methods, there are still relatively few models that explicitly incorporate them, suggesting there is still much room for model improvement.

Fourth, many models still focus only on parts of the barrier system, without considering all relevant processes. Such scientific focus up to this point was likely necessary to better understand specific system components; however, our current knowledge of important processes should lead to more complex, coupled, and fully representative models. For example, recent models, (e.g., Lorenzo-Trueba and Mariotti 2017; Walters et al., 2014), have shown the importance of coupling the backbarrier marsh–lagoon system to barrier evolution models; however, there are still relatively few models that incorporate these as coupled systems. Barrier subsidence has received relatively little attention in the literature and has been incorporated into a minority of barrier evolution models, (e.g., Rosati et al., 2006, 2010). Yet, from these few studies, we see that consolidation rates can significantly impact the future evolution of the system. The role of aeolian transport has also largely been neglected in barrier island evolution models. Although a large body of work exists regarding aeolian transport and its role in dune recovery, (e.g., Brodie *et al.*, 2017), few full-scale barrier evolution models have integrated this research. This may be the case, at least in part, because of the relatively recent focus on modeling storm sequences and post-storm beach and dune recovery (Eichentopf *et al.*, 2019). However, as various studies have indicated the importance of these morphological components, modeling efforts would be most beneficial by driving toward the incorporation of all relevant processes.

Finally, anthropogenic influences, such as urban development and its associated infrastructure, have changed and will continue to change the way many of the fundamental processes discussed in this review affect barrier island morphology. This also includes coastal engineering infrastructure, which is often intended to reduce inundation and erosion, or to support recreational and commercial navigation. Thus, modeling paradigms shifted toward representing barrier islands as coupled human–natural systems would provide important insights (McNamara and Lazarus, 2018). Modeling frameworks that included anthropogenic impacts such as the effects of human agents, (e.g., McNamara and Werner, 2008), urban development, (e.g., Rogers *et al.*, 2015), and coastal restoration practices, (Long *et al.*, 2020), would help us explore and evaluate their impacts which would be useful in coastal planning.

4.4. *Summary*

In closing, future research and development in the area of morphodynamic modeling of coastal barrier systems would benefit by leveraging existing and new datasets, advancements in observation technologies, and emerging data science approaches to better characterize morphological response and its uncertainty. Continuing the research community shift toward open-access models and data would facilitate more rapid advancement in this area. Scientific advances are most needed in understanding anthropogenic and ecological influences on barrier morphological change. Also essential is advancing scientific understanding of observed morphological phenomena and the underlying sediment transport processes, including the coupling between a barrier and its sub-systems.

Such advancements will bring us closer to achieving the overarching goal of high-confidence predictions of barrier system morphodynamics in multiple spatiotemporal dimensions.

Acknowledgments and Disclaimers

The work is reprinted with minor modifications and permission under Creative Commons Attribution 4.0 International license from *J. Waterway, Port, Coastal, and Ocean Engineering*, Advances in Morphodynamic Modeling of Coastal Barriers: A Review, Hoagland, S.W.H., Jefferies, C.R., Irish, J.L., Weiss, R., Mandli, K., Vitousek, S., Johnson, C.M., and Cialone, M.A. (2023), DOI: 10.1061/JWPED5/ WWENG-1825.

This material is based upon work that is primarily supported by the U.S. Army Corps of Engineers through the U.S. Coastal Research Program (under Grant No. W912HZ-20-2-0005) and partially supported by the National Science Foundation (under Grant Number 1735139). Any opinions, findings, and conclusions or recommendations expressed in this material are those of the authors and do not necessarily reflect the views of these organizations. This publication was also prepared in part by Steven Hoagland using Federal funds under award NA18OAR4170083, Virginia Sea Grant College Program Project R/72155T, from the National Oceanic and Atmospheric Administration's (NOAA) National Sea Grant College Program, U.S. Department of Commerce. The statements, findings, conclusions, and recommendations are those of the author(s) and do not necessarily reflect the views of Virginia Sea Grant, NOAA, or the U.S. Department of Commerce.

Disclaimer for non-endorsement of commercial products and services: Any use of trade, firm, or product names is for descriptive purposes only and does not imply endorsement by the U.S. Government.

References

Aagaard, T. and Hughes, M. (2013). Sediment transport. In J. Shroder and D. Sherman (Eds.), *Treatise on Geomorphology*, Vol. 10, Chapter 10.4. Academic Press, San Diego, CA, pp. 74–105.

Aagaard, T. and Kroon, A. (2017). Sediment transport under storm conditions on sandy beaches. In Ciavola, P. and Coco, G. (Eds.), *Coastal Storms: Processes and Impacts*, Chapter 3. John Wiley & Sons Ltd., pp. 44–63.

Alfageme, S. and Cañizares, R. (2005). Process-based morphological modeling of a restored barrier island: Whiskey Island, Louisiana, USA. In *Coastal Dynamics 2005*. American Society of Civil Engineers, Reston, VA, pp. 1–11.

Alfageme, S. R., Khondker, M., and Canizares, R. (2007). Breach stability and growth analysis using a morphological model. In *Coastal Sediments'07*. American Society of Civil Engineers, Reston, VA, pp. 2025–2036. DOI: 10.1061/40926(239)159.

Alvarez-Cuesta, M., Toimil, A., and Losada, I. J. (2021). Modelling long-term shoreline evolution in highly anthropized coastal areas. Part 1: Model description and validation. *Coastal Engineering*, 169. DOI: 10.1016/j.coastaleng.2021.103960.

Antolínez, J. A., Méndez, F. J., Anderson, D., Ruggiero, P., and Kaminsky, G. M. (2019). Predicting climate-driven coastlines with a simple and efficient multiscale model. *Journal of Geophysical Research: Earth Surface*, 124(6), 1596–1624. DOI: 10.1029/2018JF004790.

Armon, J. W. and McCann, S. B. (1979). Morphology and landward sediment transfer in a transgressive barrier island system, Southern Gulf of St. Lawrence, Canada. *Marine Geology*, 31(3–4), 333–344. DOI: 10.1016/0025-3227(79)90041-0.

Ashton, A. D. and Lorenzo-Trueba, J. (2015). Complex responses of barriers to sea-level rise emerging from a model of alongshore-coupled dynamic profile evolution. In *The Proceedings of the Coastal Sediments 2015*. World Scientific, pp. 1–7. DOI: 10.1142/9789814689977_0003.

Ashton, A. D. and Murray, A. B. (2006). High-angle wave instability and emergent shoreline shapes: 1. Modeling of sand waves, flying spits, and capes. *Journal of Geophysical Research: Earth Surface*, 111(4), 1–19. DOI: 10.1029/ 2005JF000422.

Ashton, A. D., Murray, A. B., and Amoult, O. (2001). Formation of coastline features by large-scale instabilities induced by high-angle waves. *Nature*, 414, 296–300. DOI: 10.1038/415666a.

Atkinson, A. L., Baldock, T. E., Birrien, F., Callaghan, D. P., Nielsen, P., Beuzen, T., Turner, I. L., Blenkinsopp, C. E., and Ranasinghe, R. (2018). Laboratory investigation of the Bruun rule and beach response to sea level rise. *Coastal Engineering*, 136, 183–202. DOI: 10.1016/j.coastaleng.2018.03.003.

Ayat, B. and Kobayashi, N. (2015). Vertical cylinder density and toppling effects on dune erosion and overwash. *Journal of Waterway, Port, Coastal, and Ocean Engineering*, 141(1), 04014026. DOI: 10.1061/(asce)ww.1943-5460.0000264.

Bagnold, R. A. (1963). Mechanics of marine sedimentation. In Hill, M. (Ed.), *The Sea: Ideas and Observations of Progress in the Study of the Seas, Vol. 3, The Earth Beneath the Sea; History*. John Wiley & Sons, Inc., New York, pp. 507–528.

Bagnold, R. A. (1966). An approach to the sediment transport problem from general physics. Technical Report, United States Government Printing Office, Washington, D.C.

Bakker, W. (1968). The dynamics of a coast with a groyne system. In *Proceedings of 11th ICCE*. American Society of Civil Engineers, Reston, VA, pp. 492–517. DOI: 10.31826/9781463212209-031.

Baldock, T. E., Hughes, M. G., Day, K., and Louys, J. (2005). Swash overtopping and sediment overwash on a truncated beach. *Coastal Engineering*, 52(7), 633–645. DOI: 10.1016/j.coastaleng.2005.04.002.

Banno, M., Nakamura, S., Kosako, T., Nakagawa, Y., Yanagishima, S. I., and Kuriyama, Y. (2020). Long-term observations of beach variability at Hasaki, Japan. *Journal of Marine Science and Engineering*, 8(11), 1–17. DOI: 10. 3390/jmse8110871.

Barnard, P. L., Hansen, J. E., and Erikson, L. H. (2012). Synthesis study of an erosion hot spot, Ocean Beach, California. *Journal of Coastal Research*, 28(4), 903–922. DOI: 10.2112/JCOASTRES-D-11-00212.1.

Basco, D. R. and Shin, C. S. (1999). A one-dimensional numerical model for storm-breaching of barrier islands. *Journal of Coastal Research*, 15(1), 241–260.

Belknap, D. F. and Kraft, J. C. (1985). Influence of antecedent geology on stratigraphic preservation potential and evolution of Delaware's barrier systems. *Marine Geology*, 63(1–4), 235–262. DOI: 10.1016/0025-3227(85)90085-4.

Berman, J. W., Wartman, J., Olsen, M., Irish, J. L., Miles, S. B., Tanner, T., Gurley, K., Lowes, L., Bostrom, A., Dafni, J., Grilliot, M., Lyda, A., and Peltier, J. (2020). Natural hazards reconnaissance with the NHERI RAPID facility, *Frontiers in Built Environment*, 6, 1–16. DOI: 10.3389/fbuil. 2020.573067.

Bosboom, J. and Stive, M. J. F. (2021). *Coastal Dynamics*. Delft University of Technology, Delft, The Netherlands. DOI: 10.5074/T.2021.001.

Brennan, P. (2016). Image of Beach Erosion from Hurricane Matthew. https:// pixabay.com/users/paulbr75-2938186/.

Brenner, O. T., Moore, L. J., and Murray, A. B. (2015). The complex influences of back-barrier deposition, substrate slope and underlying stratigraphy in barrier island response to sea-level rise: Insights from the Virginia Barrier Islands, Mid-Atlantic Bight, U.S.A. *Geomorphology*, 246, 334–350. DOI: 10.1016/j.geomorph.2015.06.014.

Bridges, T., King, J., Simm, J., Beck, M., Collins, G., Lodder, Q., and Mohan, R. (2021). International guidelines on natural and nature-based features for flood risk management. Technical Report, U.S. Army Corps of Engineers, Washington, D.C.

Bridges, T. S., Henn, R., Komlos, S., Scerno, D., Wamsley, T., and White, K. (2013). Coastal risk reduction and resilience. Technical Report July, United

States Army Corps of Engineers, Civil Works Directorate, Washington, D.C. http://www.swg.usace.army.mil/Portals/26/docs/PAO/Coastal.pdf.

Brodie, K. L., Palmsten, M. L., and Spore, N. J. (2017). Coastal foredune evolution, Part 1: Environmental factors & forcing processes affecting morphological evolution. Technical Report, US Army Corps of Engineers.

Bruun, P. (1962). Sea level rise as a cause of shore erosion. *Proceedings of ASCE Journal of the Waterways and Harbors Division*, 88, 117–130.

Bruun, P. (1983). Review of conditions for uses of the Bruun rule of erosion, *Coastal Engineering*, 7(1), 77–89. DOI: 10.1016/0378-3839(83)90028-5.

Buijsman, M. C. (1997). The impact of gas extraction and sea level rise on the morphology of the Wadden Sea: Extension and application of the model ASMITA. Technical Report, Netherlands Center for Coastal Research.

Callaghan, D. P., Ranasinghe, R., and Roelvink, D. (2013). Probabilistic estimation of storm erosion using analytical, semi-empirical, and process based storm erosion models. *Coastal Engineering*, 82, 64–75. DOI: 10.1016/j.coastaleng.2013.08.007.

Cañizares, R. and Irish, J. L. (2008). Simulation of storm-induced barrier island morphodynamics and flooding. *Coastal Engineering*, 55(12), 1089–1101. DOI: 10.1016/j.coastaleng.2008.04.006.

Castelle, B., Bujan, S., Marieu, V., and Ferreira, S. (2020). 16 years of topographic surveys of rip-channelled high-energy meso-macrotidal sandy beach. *Scientific Data*, 7(1), 1–9. DOI: 10.1038/s41597-020-00750-5.

Chardón-Maldonado, P., Pintado-Patiño, J. C., and Puleo, J. A. (2016). Advances in swash-zone research: Small-scale hydrodynamic and sediment transport processes. *Coastal Engineering*, 115, 8–25. DOI: 10.1016/j.coastaleng.2015.10.008.

Cho, M., Yoon, H.-D., Do, K., Son, S., and Kim, I.-H. (2019). Comparative study on the numerical simulation of bathymetric changes under storm condition. *Journal of Coastal Research*, 91(sp1), 106. DOI: 10.2112/SI91-022.1.

Ciavola, P. and Coco, G. (Eds.) (2017). *Coastal Storms*. John Wiley & Sons, Ltd., Chichester, UK. DOI: 10.1002/9781118937099.

Cleary, W. J. and Hosier, P. E. (1979). Geomorphology, washover history, and inlet zonation: Cape lookout, North Carolina to Bird Island, North Carolina. In Leatherman, S. P. (Ed.), *Barrier Islands: From the Gulf of St. Lawrence to the Gulf of Mexico*. Academic Press.

Conley, D. C. (1999). Observations on the impact of a developing inlet in a bar built estuary. *Continental Shelf Research*, 19(13), 1733–1754. DOI: 10.1016/S0278-4343(99)00035-7.

Cowell, P. J., Stive, M. J., Niedoroda, A. W., De Vriend, H. J., Swift, D. J., Kaminsky, G. M., and Capobianco, M. (2003). The coastal-tract (Part 1): A conceptual approach to aggregated modeling of low-order coastal change. *Journal of Coastal Research*, 19(4), 812–827.

Cowell, P. J., Stive, M. J., Niedoroda, A. W., Swift, D. J., De Vriend, H. J., Buijsman, M. C., Nicholls, R. J., Roy, P. S., Kaminsky, G. M., Cleveringa, J., Reed, C. W., and De Boer, P. L. (2003). The coastal-tract (Part 2): Applications of aggregated modeling of lower-order coastal change. *Journal of Coastal Research*, 19(4), 828–848.

Cowell, P. J., Thom, B. G., Jones, R. A., Everts, C. H., and Simanovic, D. (2006). Management of uncertainty in predicting climate-change impacts on beaches. *Journal of Coastal Research*, 22(1), 232–245. DOI: 10.2112/05A-0018.1.

Cowell, P., Roy, P., and Jones, R. (1992). Shoreface translation model: Computer simulation of coastal-sandbody response to sea level rise. *Mathematics and Computers in Simulation*, 33, 603–608.

Cowell, P., Roy, P., and Jones, R. (1995). Simulation of large-scale coastal change using a morphological behaviour model. *Marine Geology*, 126(1–4), 45–61. DOI: 10.1016/0025-3227(95)00065-7.

D'anna, M., Idier, D., Castelle, B., Vitousek, S., and Le Cozannet, G. (2021). Le Cozannet, Reinterpreting the bruun rule in the context of equilibrium shoreline models. *Journal of Marine Science and Engineering*, 9(9). DOI: 10.3390/jmse9090974.

Davidson, M. A., Splinter, K. D., and Turner, I. L. (2013). A simple equilibrium model for predicting shoreline change. *Coastal Engineering*, 73, 191–202. DOI: 10.1016/j.coastaleng.2012.11.002.

Davidson, M. A., Turner, I. L., Splinter, K. D., and Harley, M. D. (2017). Annual prediction of shoreline erosion and subsequent recovery. *Coastal Engineering*, 130, 14–25. DOI: 10.1016/j.coastaleng.2017.09.008.

Davies, A. G., Van Rijn, L. C., Damgaard, J. S., Van De Graaff, J., and Ribberink, J. S. (2002). Intercomparison of research and practical sand transport models, *Coastal Engineering*, 46(1), 1–23. DOI: 10.1016/S0378-3839(02)00042-X.

De Goede, E. D. (2020). Historical overview of 2D and 3D hydrodynamic modelling of shallow water flows in the Netherlands. *Ocean Dynamics*, 70(4), 521–539. DOI: 10.1007/s10236-019-01336-5.

de Schipper, M. A., de Vries, S., Ruessink, G., de Zeeuw, R. C., Rutten, J., van Gelder-Maas, C., and Stive, M. J. (2016). Initial spreading of a mega feeder nourishment: Observations of the Sand Engine pilot project. *Coastal Engineering*, 111, 23–38. DOI: 10.1016/j.coastaleng.2015.10.011.

De Vet, P. (2014). Modelling sediment transport and morphology during overwash and breaching events. PhD Thesis, Delft University of Technology.

De Vriend, H. J. (1991). Mathematical modelling and large-scale coastal behaviour - Part 1: Physical processes. *Journal of Hydraulic Research*, 29(6), 727–740. DOI: 10.1080/00221689109498955.

De Vriend, H. J. (1991). Mathematical modelling and large-scale coastal behavior - Part 2: Predictive models. *Journal of Hydraulic Research*, 29(6), 741–753. DOI: 10.1080/00221689109498956.

De Vriend, H. J., Capobianco, M., Chesher, T., de Swart, H. E., Latteux, B., and Stive, M. J. (1993). Approaches to long-term modelling of coastal morphology: A review. *Coastal Engineering*, 21(1–3), 225–269. DOI: 10.1016/0378-3839(93)90051-9.

Dean, R. G. (1991). Equilibrium beach profiles: Characteristics and applications. *Journal of Coastal Research*, 7(1), 53–84.

Dean, R. G. and Dalrymple, R. A. (2002). *Coastal Processes with Engineering Applications*. Cambridge University Press.

Dean, R. G. and Houston, J. R. (2016). Determining shoreline response to sea level rise. *Coastal Engineering*, 114, 1–8. DOI: 10.1016/j.coastaleng.2016.03.009.

Dean, R. G. and Maurmeyer, E. (1983). Models for beach profile response. In Komar, P. D. (Ed.), *Handbook of Coastal Processes and Erosion*, Chapter 7. CRC Press, Taylor and Francis Group, pp. 151–166.

Deltares. UNIBEST-CL+ (2021). https://www.deltares.nl/en/software/unibest-cl/#technical-specifications.

Dillon, W. (1970). Submergence effects on a Rhode Island barrier and lagoon and inferences on migration of barriers. *The Journal of Geology*, 78(1), 94–106.

Dissanayake, P., Brown, J., and Karunarathna, H. (2015). Impacts of storm chronology on the morphological changes of the Formby beach and dune system, UK. *Natural Hazards and Earth System Sciences*, 15(7), 1533–1543. DOI: 10.5194/nhess-15-1533-2015.

Dolan, R. and Lins, H. F. (1986). *The Outer Banks of North Carolina*. U.S. Geological Survey.

Donnelly, C., Kraus, N., and Larson, M. (2006). State of knowledge on measurement and modeling of coastal overwash. *Journal of Coastal Research*, 22(4), 965–991. DOI: 10.2112/04-0431.1.

Donnelly, C., Larson, M., and Hanson, H. (2009). A numerical model of coastal overwash. *Proceedings of the Institution of Civil Engineers - Maritime Engineering*, 162(3), 105–114. DOI: 10.1680/maen.2009.162.3.105.

Donnelly, C., Ranasinghe, R., and Larson, M. (2005). Numerical modeling of beach profile change caused by overwash. *Coastal Dynamics 2005 - Proceedings of the 5th Coastal Dynamics International Conference*, pp. 1–15. DOI: 10.1061/40855(214)56.

Dubois, R. N. (1993). Discussion of Orrin Pilkey, Robert S. Young, Stanley R. Riggs, A. W. Sam Smith, Huiyan Wu and Walter D. Pilkey, 1993. The concept of shoreface profile of equilibrium: A critical review. *Journal of Coastal Research*, 9(4), 28–31.

Edelman, T. (1968). Dune erosion during storm conditions. In *Proceedings of 11th ICCE*. American Society of Civil Engineers, Reston, VA, pp. 719–722.

Edelman, T. (1972). Dune erosion during storm conditions. In *Proceedings of 13th ICCE*. American Society of Civil Engineers, Reston, VA, pp. 1305–1311. DOI: 10.1525/9780520948068-073.

Eichentopf, S., Karunarathna, H., and Alsina, J. M. (2019). Morphodynamics of sandy beaches under the influence of storm sequences: Current research status and future needs. *Water Science and Engineering*, 12(3), 221–234. DOI: 10.1016/j.wse.2019.09.007.

Elsayed, S. M. and Oumeraci, H. (2017). Effect of beach slope and grain-stabilization on coastal sediment transport: An attempt to overcome the erosion over-estimation by XBeach. *Coastal Engineering*, 121, 179–196. DOI: 10.1016/j.coastaleng.2016.12.009.

ESA. Sentinel-2 satellite imagery (Courtesy of the U.S. Geological Survey) (2021). https://sentinel.esa.int/web/sentinel/missions/sentinel-2.

Everts, C. H. (1985). Sea level rise effects on shoreline position. *Journal of Waterway, Port, Coastal, and Ocean Engineering*, 111(6), 985–999.

Everts, C. H. (1987). Continental shelf evolution in response to a rise in sea level. In *Sea-Level Fluctuations and Coastal Evolution*. SEPM (Society for Sedimentary Geology), pp. 49–57. DOI: 10.2110/pec.87.41.0049.

Fisher, J. and Overton, M. (1984). Numerical model for dune erosion due to wave uprush. In Edge, B. L. (Ed.), *Proceedings of 19th ICCE*, American Society of Civil Engineers, Reston, VA, pp. 1553–1558.

Fisher, J. J. and Simpson, E. J. (1979). Washover and tidal sedimentation rates as environmental factors in development of a transgressive barrier shoreline. In Leatherman, S. P. (Ed.), *Barrier Islands: From the Gulf of St. Lawrence to the Gulf of Mexico*. Academic Press, pp. 127–148.

Fisher, J. S., Overton, M. F., and Chisholm, T. (1987). Field measurements of dune erosion. *Proceedings of the Coastal Engineering Conference*, 2(1984), 1107–1115. DOI: 10.1061/9780872626003.082.

Frey, A. E., Connell, K. J., Hanson, H., Larson, M., Thomas, R. C., Munger, S., and Zundel, A. (2012). GenCade version 1 model theory and user's guide. Technical Report December, US Army Corps of Engineers, Coastal and Hydraulics Laboratory.

Gerwing, T. G., Plate, E., Kidd, J., Sinclair, J., Burns, C. W., Johnson, S., Roias, S., McCulloch, C., and Bocking, R. C. (2020). Bocking. Immediate response of fish communities and water chemistry to causeway breaching and bridge installation in the Kaouk River estuary, British Columbia, Canada. *Restoration Ecology*, 28(3), 623–631. DOI: 10.1111/rec.13110.

Godfrey, P. J. (1970). Oceanic overwash and its ecological implications on the outer banks of North Carolina. Technical Report, Office of Natural Science Studies, National Parks Service, Washington, D.C.

Goldstein, E. B., Coco, G., and Plant, N. G. (2019). A review of machine learning applications to coastal sediment transport and morphodynamics. *Earth-Science Reviews*, 194, 97–108. DOI: 10.1016/j.earscirev.2019. 04.022.

Gravens, M. B., Males, R. M., and Moser, D. A. (2007). Beach-fx: Monte Carlo life- cycle simulation model for estimating shore protection project evolution and cost benefit analyses. *Shore and Beach*, 75(1), 12–19.

Grzegorzewski, A. S., Cialone, M. A., and Wamsley, T. V. (2011). Interaction of barrier islands and storms: Implications for flood risk reduction in louisiana and mississippi. *Journal of Coastal Research*, 59, 156–164. DOI: 10. 2112/ si59-016.1.

Grzegorzewski, A. S., Johnson, B. D., Wamsley, T. V., and Rosati, J. D. (2013). Sediment transport and morphology modeling of Ship Island, Mississippi, USA, during storm events. In *Coastal Dynamics 2013*. Arcachon, France, pp. 1505–1516.

Gutierrez, B. T., Williams, S. J., and Thieler, E. R. (2007). Potential for shoreline changes due to sea-level rise along the US Mid-Atlantic Region. Technical Report, U.S. Geological Survey, Reston, VA.

Hallermeier, R. J. (1980). A profile zonation for seasonal sand beaches from wave climate. *Coastal Engineering*, 4(C), 253–277. DOI: 10.1016/0378-3839 (80)90022-8.

Halsey, S. D. (1979). Nexus: New model of barrier island development. In Leatherman, S. P. (Ed.), *Barrier Islands: From the Gulf of St. Lawrence to the Gulf of Mexico*. Academic Press.

Hanson, H. and Kraus, N. C. (1989). GENESIS: Generalized model for simulating shoreline change. Technical Report CERC-89-19. Technical Report, US Army Corps of Engineers.

Harley, M. D., Kinsela, M. A., Sánchez-García, E., and Vos, K. (2019). Shoreline change mapping using crowd-sourced smartphone images. *Coastal Engineering*, 150, 175–189. DOI: 10.1016/j.coastaleng.2019.04.003.

Harter, C. and Figlus, J. (2017). Numerical modeling of the morphodynamic response of a low-lying barrier island beach and foredune system inundated during Hurricane Ike using XBeach and CSHORE. *Coastal Engineering*, 120, 64–74. DOI: 10.1016/j.coastaleng.2016.11.005.

Hoonhout, B. (2009). Dune erosion along curved coastlines. PhD Thesis, Delft University of Technology.

Houston, J. R. (1996). Discussion of: Young, R.S.; Pilkey, O.H.; Bush, D.M.; and Thieler, E.R. A discussion of the generalized model for simulating shoreline change (GENESIS). *Journal of Coastal Research*, 11(3), 875–886. *Journal of Coastal Research*, 12(4), 1038–1043.

Hoyt, J. (1969). Chenier versus barrier, genetic and stratigraphic distinction. *American Association of Petroleum Geologists Bulletin*, 53(2), 299–306.

Ibaceta, R., Splinter, K. D., Harley, M. D., and Turner, I. L. (2020). Enhanced coastal shoreline modeling using an ensemble Kalman filter to include non-stationarity in future wave climates. *Geophysical Research Letters*, 47, 1–12. DOI: 10.1029/2020GL090724.

Jiménez, J. A. and Sánchez-Arcilla, A. (2004). A long-term (decadal scale) evolution model for microtidal barrier systems. *Coastal Engineering*, 51(8–9), 749–764. DOI: 10.1016/j.coastaleng.2004.07.007.

Johnson, B. D., Kobayashi, N., and Gravens, M. B. (2012). Cross-shore numerical model CSHORE for waves, currents, sediment transport and beach profile evolution. Technical Report, US Army Corps of Engineers, Engineer Research and Development Center.

Johnson, C. L., Chen, Q., Ozdemir, C. E., Xu, K., McCall, R., and Nederhoff, K. (2021). Nederhoff, Morphodynamic modeling of a low-lying barrier subject to hurricane forcing: The role of backbarrier wetlands. *Coastal Engineering*, 167, 103886. DOI: 10.1016/j.coastaleng.2021.103886.

Johnson, D. W. (1919). *Shore Processes and Shoreline Development*. John Wiley & Sons, Inc.

Kamphuis, J. W. (1991). Alongshore sediment transport rate distribution. *Coastal Sediments '91*, 117(6), 170–183.

Kaveh, K., ReisenbÜchler, M., Lamichhane, S., Liepert, T., Nguyen, N. D., Bui, M. D., and Rutschmann, P. (2019). A comparative study of comprehensive modeling systems for sediment transport in a curved open channel. *Water*, 11. DOI: 10.3390/w11091779.

Kaveh, K., ReisenbÜchler, M., Lamichhane, S., Liepert, T., Nguyen, N. D., Bui, M. D., and Rutschmann, P. (2017). Storm clustering and beach response. In Ciavola, P. and Coco, G. (Eds.), *Coastal Storms: Processes and Impacts*, Chapter 8. John Wiley & Sons Ltd.

Keeler, A. G., McNamara, D. E., and Irish, J. L. (2018). Responding to sea level rise: Does short-term risk reduction inhibit successful long-term adaptation? *Earth's Future*, 6(4), 618–621. DOI: 10.1002/2018EF000828.

Kim, S.-c., Styles, R., Rosati, J., Ding, Y., and Permenter, R. (2020). A comparison of GenCade, Pelnard-Considere, and LITPACK. Technical Report April, United States Army Corps of Engineers.

Kobayashi, N. (1987). Analytical solution for dune erosion by storms. *Journal of Waterway, Port, Coastal, and Ocean Engineering*, 113(4), 401–418. DOI: 10.1061/(asce)0733-950x(1987)113:4(401).

Kobayashi, N. (2016). Coastal sediment transport modeling for engineering applications. *Journal of Waterway, Port, Coastal, and Ocean Engineering*, 142(6), 03116001. DOI: 10.1061/(asce)ww.1943-5460.0000347.

Kobayashi, N. and Farhadzadeh, A. (2008). Cross-shore numerical model cshore for waves, currents, sediment transport and beach profile evolution. Technical Report, Center for Applied Coastal Research, Ocean Engineering Laboratory, University of Delaware, Newark, Delaware.

Kobayashi, N., Farhadzadeh, A., Melby, J., Johnson, B. D., and Gravens, M. B. (2010). Wave overtopping of levees and overwash of dunes. *Journal of Coastal Research*, 26(5), 888–900. DOI: 10.2112/JCOASTRES-D-09-00034.1.

Kobayashi, N., Payo, A., and Schmied, L. (2008). Cross-shore suspended sand and bed load transport on beaches. *Journal of Geophysical Research: Oceans*, 113(7), 1–17. DOI: 10.1029/2007JC004203.

Kobayashi, N., Tega, Y., and Hancock, M. W. (1996). Wave reflection and overwash of dunes. *Journal of Waterway, Port, Coastal, and Ocean Engineering*, 122(3), 150–153. DOI: 10.1061/(ASCE)0733-950X(1996)122:3 (150).

Komar, P. D. and Moore, J. R. (1983). *Handbook of Coastal Processes and Erosion*, 2018 (Orig edn.). CRC Press, Taylor and Francis Group. DOI: 10.1201/9781351072908.

Komar, P. D., McDougal, W., Marra, J., and Ruggiero, P. (1999). The rational analysis of setback distances: Applications to the Oregon coast. *Shore & Beach*, 67(1), 41–49.

Kraus, N. C. (2003). Analytical model of incipient breaching of coastal barriers, *Coastal Engineering Journal*, 45(4), 511–531. DOI: 10.1142/S057856340300097X.

Kraus, N. C. and Hayashi, K. (2005). Numerical morphologic model of barrier island breaching. In *Proceedings of 29th ICCE*. World Scientific Press, pp. 2120–2132.

Kraus, N. C. and Wamsley, T. V. (2003). Coastal barrier breaching, Part 1: Overview of breaching processes. *US Army Corps of Engineers*. 1–14.

Kraus, N. C., Patsch, K., and Munger, S. (2008). Barrier beach breaching from the lagoon side, with reference to Northern California. *Shore and Beach*, 76(2), 33–43.

Kraus, N., Militello, A., and Todoroff, G. (2002). Barrier beaching processes and barrier spit breach, Stone Lagoon, California. *Shore and Beach*, 70(4), 21–28.

Kriebel, D. L. and Dean, R. G. (1985). Numerical simulation of time-dependent beach and dune erosion. *Coastal Engineering*, 9, 221–245.

Kriebel, D. L. and Dean, R. G. (1993). Convolution method for time-dependent beach-profile response. *Journal of Waterway, Port, Coastal, and Ocean Engineering*, 119(2), 204–226.

Larson, M. and Kraus, N. (1989). SBEACH: Numerical model for simulating storm-induced beach change, Technical Report CERC-89-9. Technical Report, Coastal Engineering Research Center, US Army Corps of Engineers, Vicksburg, MS.

Larson, M. and Kraus, N. C. (1994). Temporal and spatial scales of beach profile change, Duck, North Carolina. *Marine Geology*, 117(1–4), 75–94. DOI: 10.1016/0025-3227(94)90007-8.

Larson, M., Donnelly, C., Jiménez, J. A., and Hanson, H. (2009). Analytical model of beach erosion and overwash during storms. *Proceedings of the Institution of Civil Engineers - Maritime Engineering*, 162(3), 115–125. DOI: 10.1680/maen.2009.162.3.115.

Larson, M., Erikson, L., and Hanson, H. (2004). An analytical model to predict dune erosion due to wave impact. *Coastal Engineering*, 51(8–9), 675–696. DOI: 10.1016/j.coastaleng.2004.07.003.

Larson, M., Hanson, H., and Kraus, N. C. (1987). Analytical solutions of the one-line model of shoreline change, Technical Report CERC-87-15.

Larson, M., Kraus, N. C., and Hanson, H. (1990). Decoupled numerical model of three-dimensional beach change. In *Proceedings of 22nd ICCE*. American Society of Civil Engineers, Delft, The Netherlands.

Larson, M., Kraus, N. C., and Hanson, H. (2002). Simulation of regional long-shore sediment transport and coastal evolution - The "cascade" model. In *Coastal Engineering 2002*. World Scientific Publishing Company, pp. 2612–2624. DOI: 10.1142/9789812791306_0218.

Larson, M., Palalane, J., Fredriksson, C., and Hanson, H. (2016). Simulating cross-shore material exchange at decadal scale. Theory and model component validation. *Coastal Engineering*, 116, 57–66. DOI: 10.1016/j.coastaleng.2016.05.009.

Larson, M., Wise, R., and Kraus, N. C. (2004). Coastal overwash, Part 2: Upgrade to SBEACH, Technical Note ERDC/CHL CHETN-XIV-14. Technical Report September, US Army Engineer Research and Development Center, Coastal and Hydraulics Laboratory, Vicksburg, MS.

Latteux, B. (1995). Techniques for long-term morphological simulation under tidal action. *Marine Geology*, 126(1–4), 129–141. DOI: 10.1016/ 0025-3227(95)00069-B.

Leatherman, S. P. (1976). Quantification of overwash processes. PhD Thesis, University of Virginia, Charlottesville, Virginia.

Leatherman, S. P. (1979). Migration of Assateague Island, Maryland, by inlet and overwash processes. *Geology*, 7, 104–107.

Leatherman, S. P. (1983). Barrier dynamics and landward migration with Holocene sea-level rise. *Nature*, 301(3), 415–417.

Leatherman, S. P. (1987). Annotated chronological bibliography of Barrier Island migration. *Journal of Coastal Research*, 3(1), 1 14.

Lentz, E. E. and Hapke, C. J. (2011). Geologic framework influences on the geomorphology of an anthropogenically modified barrier island: Assessment of dune/beach changes at Fire Island, New York. *Geomorphology*, 126(1–2), 82–96. DOI: 10.1016/j.geomorph.2010.10.032.

Lesser, G. R., Roelvink, J. A., van Kester, J. A., and Stelling, G. S. (2004). Development and validation of a three-dimensional morphological model. *Coastal Engineering*, 51(8–9), 883–915. DOI: 10.1016/j.coastaleng.2004.07.014.

List, J. H., Farris, A. S., and Sullivan, C. (2006). Reversing storm hotspots on sandy beaches: Spatial and temporal characteristics. *Marine Geology*, 226, 261–279. DOI: 10.1016/j.margeo.2005.10.003.

Long, J. W. and Plant, N. G. (2012). Extended Kalman filter framework for forecasting shoreline evolution. *Geophysical Research Letters*, 39(13), 1–6. DOI: 10.1029/2012GL052180.

Long, J., Dalyander, P. S., Poff, M., Spears, B., Borne, B., Thompson, D., Mickey, R., Dartez, S., and Grandy, G. (2020). Event and decadal-scale modeling of barrier island restoration designs for decision support. *Shore & Beach*, 88(1), 49–57. DOI: 10.34237/1008816.

Lorenzo-Trueba, J. and Ashton, A. D. (2014). Rollover, drowning, and discontinuous retreat: Distrinct modes of barrier response to sea-level rise arising from a simple morphodynamic model. *Journal of Geophysical Research: Earth Surface*, 779–801. DOI: 10.1002/2013JF002871.

Lorenzo-Trueba, J. and Mariotti, G. (2017). Chasing boundaries and cascade effects in a coupled barrier-marsh-lagoon system. *Geomorphology*, 290, 153–163. DOI: 10.1016/j.geomorph.2017.04.019.

Ludka, B. C., Guza, R. T., O'Reilly, W. C., Merrifield, M. A., Flick, R. E., Bak, A. S., Hesser, T., Bucciarelli, R., Olfe, C., Woodward, B., Boyd, W., Smith, K., Okihiro, M., Grenzeback, R., Parry, L., and Boyd, G. (2019). Sixteen years of bathymetry and waves at San Diego beaches. *Scientific Data*, 6(1), 1–13. DOI: 10.1038/s41597-019-0167-6.

Luijendijk, A., Hagenaars, G., Ranasinghe, R., Baart, F., Donchyts, G., and Aarninkhof, S. (2018). The state of the world's beaches. *Scientific Reports*, 8(1), 6641. DOI: 10.1038/s41598-018-24630-6.

Mariotti, G. and Carr, J. (2014). Dual role of salt marsh retreat: Long-term and short-term resilience. *Water Resources Research*, 50(4), 2963–2974. DOI: 10.1002/2013WR014676.

Mariotti, G. and Fagherazzi, S. (2010). A numerical model for the coupled long-term evolution of salt marshes and tidal flats. *Journal of Geophysical Research*, 115, F01004. DOI: 10.1029/2009JF001326.

Masetti, R., Fagherazzi, S., and Montanari, A. (2008). Application of a barrier island translation model to the millennial-scale evolution of Sand Key, Florida, *Continental Shelf Research*, 28(9), 1116–1126. DOI: 10.1016/j.csr.2008.02.021.

McBride, R. A., Anderson, J. B., Buynevich, I. V., Cleary, W., Fenster, M. S., FitzGerald, D. M., Harris, M. S., Hein, C. J., Klein, A. H., Liu, B., de Menezes, J. T., Pejrup, M., Riggs, S. R., Short, A. D., Stone, G. W., Wallace, D. J., and Wang, P. (2013). Morphodynamics of barrier systems: A synthesis. In Shroder, J. F. (Ed.), *Treatise on Geomorphology*, Vol. 10. Elsevier, pp. 166–244. DOI: 10.1016/B978-0-12-374739-6.00279-7.

McCall, R. T., Van Thiel de Vries, J. S., Plant, N. G., Van Dongeren, A. R., Roelvink, J. A., Thompson, D. M., and Reniers, A. J. (2010).

Two-dimensional time dependent hurricane overwash and erosion modeling at Santa Rosa Island. *Coastal Engineering*, 57(7), 668–683. DOI: 10.1016/j.coastaleng.2010.02.006.

McCann, S. (1972). Reconnaissance survey of Hog Island, Prince Edward Island, *Maritime Sediments*, 8(3), 107–113.

McCann, S. (1979). Barrier islands in the Southern Gulf of St. Lawrence, Canada. In Leatherman, S. P. (Ed.), *Barrier Islands: From the Gulf of St. Lawrence to the Gulf of Mexico*. Academic Press.

McCarroll, R. J., Masselink, G., Valiente, N. G., Scott, T., Wiggins, M., Kirby, J. A., and Davidson, M. A. (2021). A rules-based shoreface translation and sediment budgeting tool for estimating coastal change: ShoreTrans. *Marine Geology*, 435, 106466. DOI: 10.1016/j.margeo.2021.106466.

McNamara, D. E. and Keeler, A. (2013). A coupled physical and economic model of the response of coastal real estate to climate risk. *Nature Climate Change*, 3(6), 559–562. DOI: 10.1038/nclimate1826.

McNamara, D. E. and Lazarus, E. D. (2018). Barrier islands as coupled human-lanscape systems. In Moore, L. J. and Murray, A. B. (Eds.), *Barrier Dynamics and Response to Changing Climate*. Springer, New York, pp. 363–383.

McNamara, D. E. and Werner, B. T. (2008). Coupled barrier island-resort model: 1. Emergent instabilities induced by strong human-landscape interactions. *Journal of Geophysical Research: Earth Surface*, 113(1), 1–10. DOI: 10.1029/2007JF000840.

Miller, J. K. and Dean, R. G. (2004). A simple new shoreline change model. *Coastal Engineering*, 51(7), 531–556. DOI: 10.1016/j.coastaleng.2004.05.006.

Montaño, J., Coco, G., Antolínez, J. A., Beuzen, T., Bryan, K. R., Cagigal, L., Castelle, B., Davidson, M. A., Goldstein, E. B., Ibaceta, R., Idier, D., Ludka, B. C., Masoud-Ansari, S., Méndez, F. J., Murray, A. B., Plant, N. G., Ratliff, K. M., Robinet, A., Rueda, A., Sénéchal, N., Simmons, J. A., Splinter, K. D., Stephens, S., Townend, I., Vitousek, S., and Vos, K. (2020). Blind testing of shoreline evolution models. *Scientific Reports*, 10(1), 1–10. DOI: 10.1038/s41598-020-59018-y.

Moody, D. (1964). Coastal morphology and processes in relation to the development of submarine sand ridges off Bethany Beach, Delaware. PhD Thesis, Johns Hopkins University, Baltimore, Maryland.

Moore, L. J. and Murray, A. B. (Eds.) (2018). *Barrier Dynamics and Response to Changing Climate*. Springer International Publishing, Cham, Switzerland. DOI: 10.1007/978-3-319-68086-6.

Moore, L. J., List, J. H., Williams, S. J., and Stolper, D. (2007). Modeling barrier island response to sea-level rise in the outer banks, North Carolina. In *Coastal Sediments '07*. American Society of Civil Engineers, Reston, VA, pp. 1153–1164.

Moore, L. J., List, J. H., Williams, S. J., and Stolper, D. (2010). Complexities in barrier island response to sea level rise: Insights from numerical model experiments, North Carolina Outer Banks. *Journal of Geophysical Research*, 115, F03004. DOI: 10.1029/2009jf001299.

Moslow, T. F. and Heron Jr., S. D. (1979). Quaternary evolution of core banks, North Carolina: Cape lookout to new drum inlet. In Leatherman, S. P. (Ed.), *Barrier Islands: From the Gulf of St. Lawrence to the Gulf of Mexico.* Academic Press.

Mull, J. and Ruggiero, P. (2014). Estimating storm-induced dune erosion and over-topping along U.S. West Coast Beaches. *Journal of Coastal Research*, 298(6), 1173–1187. DOI: 10.2112/jcoastres-d-13-00178.1.

Murray, A. B. (2003). Contrasting the goals, strategies, and predictions associated with simplified numerical models and detailed simulations. In Wilcock, P. R. and Iverson, R. M. (Eds.), *Prediction in Geomorphology, Geophysical Monograph 135*. American Geophysical Union, Washington, D.C., pp. 151–165. DOI: 10.1029/135GM11.

NASEM. (2018). *Understanding the Long-Term Evolution of the Coupled Natural- Human Coastal System.* The National Academies Press. DOI: 10. 17226/25108.

Nichols, R. and Marston, A. (1939). Shoreline changes in Rhode Island produced by hurricane of September 21, 1938. *Geological Society of America Bulletin*, 50.

Niedoroda, A. W., Reed, C. W., Swift, D. J., Arato, H., and Hoyanagi, K. (1995). Modeling shore-normal large-scale coastal evolution. *Marine Geology*, 126, 181–199. DOI: 10.1016/0025-3227(95)98961-7.

Nielsen, P. (2009). *Coastal and Estuarine Processes.* World Scientific Publishing Company.

Nienhuis, J. H. and Lorenzo-Trueba, J. (2019). Simulating barrier island response to sea level rise with the barrier island and inlet environment (BRIE) model v1.0. *Geoscientific Model Development*, 12(9), 4013–4030. DOI: 10.5194/ gmd-12-4013-2019.

Nienhuis, J. H., Heijkers, L. G., and Ruessink, G. (2021). Barrier breaching versus overwash deposition: Predicting the morphologic impact of storms on coastal barriers. *Journal of Geophysical Research: Earth Surface*, 126(6), 1–17 DOI: 10.1029/2021JF006066.

Nishi, R. and Kraus, N. (1996). Mechanism and calculation of sand dune erosion by storms. In *Proceedings of 26th ICCE*. American Society of Civil Engineers, Reston, VA, pp. 3034–3047.

Oertel, G. F. (1979). Barrier island development during the holocene recession, Southeastern United States. In Leatherman, S. P. (Ed.), *Barrier Islands: From the Gulf of St. Lawrence to the Gulf of Mexico.* Academic Press, pp. 273–290.

Oppenheimer, M., Glavovic, B., Hinkel, J., van de Wal, R., Magnan, A. K., Abd-Elgawad, A., Cai, R., Cifuentes-Jara, M., Deconto, R. M., Ghosh, T., Hay, J., Isla, F., Marzeion, B., Meyssignac, B., and Sebesvari, Z. (2019). Sea level rise and implications for low lying islands, coasts and communities coordinating. In H.-O. Portner, D. Roberts, V. Masson-Delmotte, P. Zhai, M. Tignor, E. Poloczanska, K. Mintenbeck, A. Alegria, M. Nicolai, A. Okem, J. Petzold, B. Rama, and N. Weye (Eds.), *IPCC SR Ocean and Cryosphere* (In press).

Otvos Jr., E. (1970). Development and migration of barrier islands, Northern Gulf of Mexico. *Geological Society of America Bulletin*, 81, 241–246.

Over, J.-S., Brown, J., Sherwood, C., Hegermiller, C., Wernette, P., Ritchie, A., and Warrick, J. (2021). A survey of storm-induced seaward-transport features observed during the 2019 and 2020 hurricane seasons. *Shore & Beach*, 89(2), 31–40. DOI: 10.34237/1008924.

Overton, M. F., Fisher, J. S., and Young, M. A. (1988). Laboratory investigation of dune erosion. *Journal of Waterway, Port, Coastal, and Ocean Engineering*, 114(3), 367–373. DOI: 10.1061/(ASCE)0733-950X(1988)114:3(367).

Overton, M. F., Pratikto, W. A., Lu, J. C., and Fisher, J. S. (1994). Laboratory investigation of dune erosion as a function of sand grain size and dune density. *Coastal Engineering*, 23(1–2), 151–165. DOI: 10.1016/0378-3839(94)90020-5.

Palalane, J. and Larson, M. (2020). A long-term coastal evolution model with long- shore and cross-shore transport. *Journal of Coastal Research*, 36(2), 411–423. DOI: 10.2112/JCOASTRES-D-17-00020.1.

Palmsten, M. L. and Holman, R. A. (2012). Laboratory investigation of dune erosion using stereo video. *Coastal Engineering*, 60(1), 123–135. DOI: 10. 1016/j.coastaleng.2011.09.003.

Passeri, D. L., Long, J. W., Plant, N. G., Bilskie, M. V., and Hagen, S. C. (2018). The influence of bed friction variability due to land cover on storm-driven barrier island morphodynamics. *Coastal Engineering*, 132, 82–94. DOI: 10.1016/j.coastaleng.2017.11.005.

Pelnard-Considere, R. (1956). Essai de theorie de l'evolution des formes de rivage en plages de sable et de galets. *Journees de L'hydraulique*, 289–298.

Pender, D. and Karunarathna, H. (2013). A statistical-process based approach for modelling beach profile variability. *Coastal Engineering*, 81, 19–29. DOI: 10.1016/j.coastaleng.2013.06.006.

Penland, S. and Boyd, R. (1981). Shoreline changes on the Louisiana Barrier coast. *Oceans*. DOI: 10.1017/CBO9781107415324.004.

Perlin, M. and Dean, R. G. (1979). Prediction of beach planforms with littoral controls. *Proceedings of the Coastal Engineering Conference*, 2(1), 1818–1838. DOI: 10.9753/icce.v16.110.

Perlin, M. and Dean, R. G. (1985). 3-D model of bathymetric response to structures. *Journal of Waterway, Port, Coastal, and Ocean Engineering*, 111(2), 153–170.

Pierce, J. (1969). Sediment budget along a barrier island chain. *Sedimentary Geology*, 3, 5–16.

Pierce, J. W. (1970). Tidal inlets and washover fans. *The Journal of Geology*, 78(2), 230–234.

Pilkey, O. H. and Davis, T. W. (1987). An analysis of coastal recession models: North Carolina coast. In *Sea-Level Fluctuation and Coastal Evolution*. SEPM (Society for Sedimentary Geology), pp. 59–68.

Pilkey, O. H., Neal, W. J., Kelley, J. T., and Cooper, J. A. G. (2011). *The World's Beaches: A Global Guide to the Science of the Shoreline*. University of California Press.

Pilkey, O. H., Young, R. S., Riggs, S. R., Smith, A. W., and Pilkey, W. D. (1993). The concept of shoreface profile of equilibrium: A critical review. *Journal of Coastal Research*, 9(1), 255–278.

Quartel, S., Ruessink, B. G., and Kroon, A. (2007). Daily to seasonal cross-shore behaviour of quasi-persistent intertidal beach morphology. *Earth Surface Processes and Landforms*, 32, 1293–1307. DOI: 10.1002/esp.1477.

Ranasinghe, R. (2020). On the need for a new generation of coastal change models for the 21st century. *Scientific Reports*, 10(1), 1–6. DOI: 10.1038/s41598-020-58376-x.

Ranasinghe, R., Callaghan, D., and Stive, M. J. (2012). Estimating coastal recession due to sea level rise: Beyond the Bruun rule. *Climatic Change*, 110(3–4), 561–574. DOI: 10.1007/s10584-011-0107-8.

Ranasinghe, R., Holman, R., De Schipper, M., Lippmann, T., Wehof, J., Duong, T. M., Roelvink, D., and Stive, M. (2012). Quantifying nearshore morphological recovery time scales using argus video imaging: Palm Beach, Sydney and Duck, North Carolina. *Proceedings of the Coastal Engineering Conference*, 1–7. DOI: 10.9753/icce.v33.sediment.24.

Rathje, E. M., Dawson, C., Padgett, J. E., Pinelli, J.-P., Stanzione, D., Adair, A., Arduino, P., Brandenberg, S. J., Cockerill, T., Dey, C., Esteva, M., Haan, F. L., Hanlon, M., Kareem, A., Lowes, L., Mock, S., and Mosqueda, G. (2017). DesignSafe: New cyberinfrastructure for natural hazards engineering. *Natural Hazards Review*, 18(3), 06017001. DOI: 10.1061/(asce)nh.1527-6996.0000246.

Raubenheimer, B. (2020). Development of a nearshore extreme events reconnaissance community. *Coastal Engineering Proceedings*. DOI: 10.9753/icce.v36v.keynote.12.

Reeve, D. E., Karunarathna, H., Pan, S., Horrillo-Caraballo, J. M., Różyñski, G., and Ranasinghe, R. (2016). Data-driven and hybrid coastal morphological prediction methods for mesoscale forecasting. *Geomorphology*, 256, 49–67. DOI: 10.1016/j.geomorph.2015.10.016.

Reeves, I. R., Moore, L. J., Murray, A. B., Anarde, K. A., and Goldstein, E. B. (2021). Dune dynamics drive discontinuous barrier retreat. *Geophysical Research Letters*, 48(13), 1–11. DOI: 10.1029/2021GL092958.

Robinet, A., Idier, D., Castelle, B., and Marieu, V. (2018). A reduced-complexity shoreline change model combining longshore and cross-shore processes: The LX-Shore model. *Environmental Modelling and Software*, 109, 1–16. DOI: 10.1016/j.envsoft.2018.08.010.

Roelvink, D., Huisman, B., Elghandour, A., Ghonim, M., and Reyns, J. (2020). Efficient modeling of complex sandy coastal evolution at monthly to century time scales. *Frontiers in Marine Science*, 7, 1–20. DOI: 10.3389/fmars.2020.00535.

Roelvink, D., Reniers, A., van Dongeren, A., van Thiel de Vries, J., McCall, R., and Lescinski, J. (2009). Modelling storm impacts on beaches, dunes and barrier islands. *Coastal Engineering*, 56(11–12), 1133–1152. DOI: 10.1016/j.coastaleng.2009.08.006.

Roelvink, J. A. (2006). Coastal morphodynamic evolution techniques. *Coastal Engineering*, 53(2–3), 277–287. DOI: 10.1016/j.coastaleng.2005.10.015.

Rogers, L. J., Moore, L. J., Goldstein, E. B., Hein, C. J., Lorenzo-Trueba, J., and Ashton, A. D. (2015). Anthropogenic controls on overwash deposition: Evidence and consequences. *Journal of Geophysical Research: Earth Surface*, 120(12), 2609–2624. DOI: 10.1002/2015JF003634.

Rosati, J. D. and Stone, G. W. (2009). Geomorphologic evolution of barrier islands along the Northern U.S. Gulf of Mexico and implications for engineering design in barrier restoration. *Journal of Coastal Research*, 251, 8–22. DOI: 10.2112/07-0934.1.

Rosati, J. D., Dean, R. G., and Stone, G. W. (2010). A cross-shore model of barrier island migration over a compressible substrate. *Marine Geology*, 271(1–2), 1–16. DOI: 10.1016/j.margeo.2010.01.005.

Rosati, J. D., Dean, R. G., and Walton, T. L. (2013). The modified Bruun Rule extended for landward transport. *Marine Geology*, 340, 71–81. DOI: 10.1016/j.margeo.2013.04.018.

Rosati, J. D., Dean, R. G., Kraus, N. C., and Stone, G. W. (2006). Morphologic evolution of subsiding barrier island systems. In *Coastal Engineering 2006*. World Scientific Publishing Company, San Diego, California, pp. 3963–3975. DOI: 10.1142/9789812709554_0333.

Rosen, P. S. (1979). Aeolian dynamics of a barrier island system. In Leatherman, S. P. (Ed.), *Barrier Islands: From the Gulf of St. Lawrence to the Gulf of Mexico*. Academic Press, pp. 81–98.

Ruessink, B. G., Kuriyama, Y., Reniers, A. J., Roelvink, J. A., and Walstra, D. J. R. (2007). Modeling cross-shore sandbar behavior on the timescale of weeks. *Journal of Geophysical Research: Earth Surface*, 112(3), 1–15. DOI: 10.1029/2006JF000730.

Ruggiero, P., Buijsman, M., Kaminsky, G. M., and Gelfenbaum, G. (2010). Modeling the effects of wave climate and sediment supply variability on large-scale shoreline change. *Marine Geology*, 273(1–4), 127–140. DOI: 10.1016/j.margeo.2010.02.008.

Safak, I., List, J. H., Warner, J. C., and Schwab, W. C. (2017). Persistent shoreline shape induced from offshore geologic framework: Effects of shoreface connected ridges. *Journal of Geophysical Research: Oceans*, 122(11), 8721–8738. DOI: 10.1002/2017JC012808.

Safak, I., Warner, J. C., and List, J. H. (2016). Barrier island breach evolution: Alongshore transport and bay-ocean pressure gradient interactions. *Journal of Geophysical Research: Oceans*, 121, 8720–8730. DOI: 10.1002/2016JC012029.

Sallenger, A. H. (2000). Storm impact scale for barrier islands. *Journal of Coastal Research*, 16(3), 890–895.

Sánchez-Arcilla, A. and Jiménez, J. A. (1994). Breaching in a wave-dominated barrier spit: The trabucador bar (North-Eastern Spanish coast). *Earth Surface Processes and Landforms*, 19(6), 483–498. DOI: 10.1002/esp.3290190602.

Schwartz, M. L. (1967). The Bruun theory of sea-level rise as a cause of shore erosion. *The Journal of Geology*, 75(1), 76–92.

Schwartz, M. L. (1973). Barrier islands. In *Benchmark Papers in Geology*. Dowden, Hutchinson & Ross, Inc.

Shaler, N. (1895). *Beaches and Tidal Marshes of the Atlantic Coast*. American Book Company.

Shaw, L., Helmholz, P., Belton, D., and Addy, N. (2019). Comparison of UAV Lidar and imagery for beach monitoring. *International Archives of the Photogrammetry, Remote Sensing and Spatial Information Sciences - ISPRS Archives*, 42(2/W13), 589–596. DOI: 10.5194/isprs-archives-XLII 2-W13-589-2019.

Shephard, F. P. (1950). Beach cycles in Southern California. Technical Report, United States Army Corps of Engineers.

Sherwood, C. R., Long, J. W., Dickhudt, P. J., Dalyander, P. S., Thompson, D. M., and Plant, N. G. (2014). Plant. Inundation of a barrier island (Chandeleur Islands, Louisiana, USA) during a hurricane: Observed water-level gradients and modeled seaward sand transport. *Journal of Geophysical Research: Earth Surface*, 119(7), 1498–1515. DOI: 10.1002/2013JF003069.

Sherwood, C. R., van Dongeren, A., Doyle, J., Hegermiller, C. A., Hsu, T.-J., Kalra, T. S., Olabarrieta, M., Penko, A. M., Rafati, Y., Roelvink, D., van der Lugt, M., Veeramony, J., and Warner, J. C. (2022). Modeling the morphodynamics of coastal responses to extreme events: What shape are we in? *Annual Review of Marine Science*, 14(1), 1–36. DOI: 10.1146/annurev-marine-032221-090215.

Shin, C. S. (1996). A one-dimensional model for storm breaching of barrier islands. PhD Thesis, Old Dominion University.

Simmons, J. A., Splinter, K. D., Harley, M. D., and Turner, I. L. (2019). Calibration data requirements for modelling subaerial beach storm erosion. *Coastal Engineering*, 152, 103507. DOI: 10.1016/j.coastaleng.2019.103507.

Smallegan, S. M. and Irish, J. L. (2017). Barrier island morphological change by bay-side storm surge. *Journal of Waterway, Port, Coastal, and Ocean Engineering*, 143(5), 04017025. DOI: 10.1061/(asce)ww.1943-5460. 0000413.

Soulsby, R. L. (1997). *Dynamics of Marine Sands*. Thomas Telford, London.

Splinter, K. D., Carley, J. T., Golshani, A., and Tomlinson, R. (2014). A relationship to describe the cumulative impact of storm clusters on beach erosion, *Coastal Engineering*, 83, 49–55. DOI: 10.1016/j.coastaleng.2013.10.001.

Steetzel, H. J. (1993). Cross-shore transport during storm surges. PhD Thesis, Technische Universiteit Delft.

Steetzel, H. J., de Vroeg, H., van Rijn, L. C., and Stam, J. M. (1998). Morphological modelling using a modified multi-layer approach. *Proceedings of the Coastal Engineering Conference*, 2, 2368–2381. DOI: 10.1061/9780784404119.178.

Stive, M. J. and de Vriend, H. J. (1995). Modelling shoreface profile evolution. *Marine Geology*, 126, 235–248. DOI: 10.1016/0025-3227(95)00080-I.

Stive, M. J., De Vriend, H. J., Cowell, P. J., and Niedoroda, A. W. (1995). Behaviour-oriented models of shoreface evolution. In *Coastal Dynamics - Proceedings of the International Conference*. ASCE, pp. 998–1005.

Stive, M. J., Roelvink, D. J., and de Vriend, H. (1990). Large-scale coastal evolution concept. In *Proceedings of 22nd ICCE*. ASCE, pp. 1962–1974.

Stolper, D., List, J. H., and Thieler, E. R. (2005). Simulating the evolution of coastal morphology and stratigraphy with a new morphological-behaviour model (GEOMBEST). *Marine Geology*, 218, 17–36. DOI: 10.1016/j. margeo.2005.02.019.

Storms, J. E., Weltje, G., van Dijke, J., Geel, C., and Kroonenberg, S. (2002). Process-response modeling of wave-dominated coastal systems: Simulating evolution and stratigraphy on geological timescales. *Journal of Sedimentary Research*, 72(2), 226–239. DOI: 10.1306/052501720226.

Stutz, M. L. and Pilkey, O. H. (2011). Open-ocean barrier islands: Global influence of climatic, oceanographic, and depositional settings. *Journal of Coastal Research*, 272, 207–222. DOI: 10.2112/09-1190.1.

Swift, D. J. (1975). Barrier-island genesis: Evidence from the central atlantic shelf, eastern U.S.A. *Sedimentary Geology*, 14(1), 1–43. DOI: 10.1016/ 0037-0738(75)90015-9.

Tanaka, H., Suntoyo, and Nagasawa, T. (2002). Sediment intrusion into Gamo Lagoon by wave overtopping. In *Coastal Engineering 2002*. World Scientific Publishing Company, pp. 823–835. DOI: 10.1142/9789812791306_0070.

Terwindt, J. and Battjes, J. (1990). Research on large-scale coastal behavior. In *Proceedings of 22nd ICCE*. ASCE, pp. 1975–1983.

Thieler, E. R., Pilkey, O. H., Young, R. S., Bush, D. M., and Chai, F. (2000). The use of mathematical models to predict beach behavior for U.S. coastal engineering: A critical review. *Journal of Coastal Research*, 16(1), 48–70.

Thomas, R. C. and Frey, A. E. (2013). Shoreline change modeling using one-line models: General model comparison and literature review. *ERDC/CHL CHETN-II-55. Vicksburg, MS: US Army Engineer Research and Development Center*, 1956. DOI: ERDC/CHLCHETN-II-55.

Toimil, A., Camus, P., Losada, I. J., Le Cozannet, G., Nicholls, R. J., Idier, D., and Maspataud, A. (2020). Climate change-driven coastal erosion modelling in temperate sandy beaches: Methods and uncertainty treatment. *Earth-Science Reviews*, 202, 103110. DOI: 10.1016/j.earscirev.2020.103110.

Toimil, A., Losada, I. J., Camus, P., and Díaz-Simal, P. (2017). Managing coastal erosion under climate change at the regional scale. *Coastal Engineering*, 128, 106–122. DOI: 10.1016/j.coastaleng.2017.08.004.

Troy, C. D., Cheng, Y. T., Lin, Y. C., and Habib, A. (2021). Rapid lake Michigan shoreline changes revealed by UAV LiDAR surveys. *Coastal Engineering*, 170, 104008. DOI: 10.1016/j.coastaleng.2021.104008.

Turner, I. L., Harley, M. D., Almar, R., and Bergsma, E. W. (2021). Satellite optical imagery in coastal engineering. *Coastal Engineering*, 167, 103919. DOI: 10.1016/j.coastaleng.2021.103919.

Turner, I. L., Harley, M. D., Short, A. D., Simmons, J. A., Bracs, M. A., Phillips, M. S., and Splinter, K. D. (2016). A multi-decade dataset of monthly beach profile surveys and inshore wave forcing at Narrabeen, Australia. *Scientific Data*, 3, 1–13. DOI: 10.1038/sdata.2016.24.

USACE. *Shore Protection Manual: Vol. I.* US Army Corps of Engineers, Coastal Engineering Research Center, Vicksburg, MS (1984).

Van Baaren, P. F. (2007). Influence of the wave period in the dune erosion model DUROSTA. PhD Thesis, Delft University of Technology.

van der Lugt, M. A., Quataert, E., van Dongeren, A., van Ormondt, M., and Sherwood, C. R. (2019). Morphodynamic modeling of the response of two barrier islands to Atlantic hurricane forcing. *Estuarine, Coastal and Shelf Science*, 229, 106404. DOI: 10.1016/j.ecss.2019.106404.

Van Dongeren, A., Bolle, A., Vousdoukas, M. I., Plomaritis, T., Eftimova, P., Williams, J., Armaroli, C., Idier, D., Van Geer, P., Van Thiel de Vries, J., Haerens, P., Taborda, R., Benavente, J., Trifonova, E., Ciavola, P., Balouin, Y., and Roelvink, D. (2009). MICORE: Dune erosion and overwash model validation with data from nine European field sites. In *Proceedings of Coastal Dynamics 2009*. World Scientific, pp. 1–15. DOI: 10.1142/9789814282475_0084.

van Ormondt, M., Nelson, T. R., Hapke, C. J., and Roelvink, D. (2020). Morphodynamic modelling of the wilderness breach, Fire Island, New York. Part I: Model set-up and validation. *Coastal Engineering*, 157, 103621. DOI: 10.1016/j.coastaleng.2019.103621.

van Rijn, L. (1993). *Principles of Sediment Transport in Rivers, Estuaries and Coastal Seas*. Aqua Publications, Amsterdam.

Van Rijn, L. C. (1984). Sediment transport: Bed load transport. *Journal of Hydraulic Engineering - ASCE*, 110(10), 1431–1456.

Vellinga, P. (1986). Beach and dune erosion during storm surges. PhD Thesis, Delft University of Technology, Delft, Netherlands.

Vemulakonda, S. R., Scheffner, N. W., Earickson, J. A., and Chou, L. W. (1988). Kings Bay coastal processes numerical model, Technical Report - US Army Coastal Engineering Research Center. 88–3.

VGIN. Virginia Base Mapping Program (VBMP) Orthoimagery (2021). https://vgin.vdem.virginia.gov/pages/base-mapping.

Visser, P. J. (1998). Breach erosion in sand-dikes. *Proceedings of the Coastal Engineering Conference*, 3, 3516–3528. DOI: 10.9753/icce.v26.

Visser, P. J. (2001). A model for breach erosion in sand-dikes. *Coastal Engineering 2000 - Proceedings of the 27th International Conference on Coastal Engineering, ICCE 2000*, Vol. 276, pp. 3829–3842. DOI: 10.1061/40549(276)299.

Vitousek, S. and Barnard, P. L. (2015). A nonlinear, implicit, one-line model to predict long-term shoreline change. In *The Proceedings of the Coastal Sediments 2015*. World Scientific, pp. 1–14. DOI: 10.1142/9789814689977_0215.

Vitousek, S., Barnard, P. L., Limber, P., Erikson, L., and Cole, B. (2017). A model integrating longshore and cross-shore processes for predicting long-term shoreline response to climate change. *Journal of Geophysical Research: Earth Surface*, 782–806. DOI: 10.1002/2016JF004065.

Vitousek, S., Cagigal, L., Montaño, J., and Rueda, A. (2021). The application of ensemble wave forcing to quantify uncertainty of shoreline change predictions journal of geophysical research: Earth surface. *Journal of Geophysical Research: Earth Surface*, 1–43.

Vos, K., Harley, M. D., Splinter, K. D., Simmons, J. A., and Turner, I. L. (2019). Sub-annual to multi-decadal shoreline variability from publicly available satellite imagery. *Coastal Engineering*, 150, 160–174. DOI: 10.1016/j.coastaleng.2019.04.004.

Walters, D., Moore, L. J., Vinent, O. D., Fagherazzi, S., and Mariotti, G. (2014). Interactions between barrier islands and backbarrier marshes affect island system response to sea level rise: Insights from a coupled model. *Journal of Geophysical Research: Earth Surface*, 119(9), 2013–2031. DOI: 10.1002/2014JF003091.

Wamsley, T. V. and Kraus, N. C. (2005). Coastal barrier island breaching, Part 2: Mechanical breaching and breach closure. *Erdc/Chl Chetn-Iv-65*. 21.

Wamsley, T. V., Cialone, M. A., Smith, J. M., Ebersole, B. A., and Grzegorzewski, A. S. (2009). Influence of landscape restoration and degradation on storm surge and waves in Southern Louisiana. *Natural Hazards*, 51(1), 207–224. DOI: 10.1007/s11069-009-9378-z.

Warner, J. C., Armstrong, B., He, R., and Zambon, J. B. (2010). Development of a coupled ocean-atmosphere-wave-sediment transport (COAWST) modeling system. *Ocean Modelling*, 35(3), 230–244. DOI: 10.1016/j.ocemod.2010.07.010.

Warner, J. C., Olabarrieta, M., Sherwood, C. R., Hegermiller, C., and Kalra, T. S. (2018). Investigations of morphological changes during hurricane sandy using a coupled modeling system. In *AGU Fall Meeting Abstracts*. American Geophysical Union, Washington, D.C.

Wartman, J., Berman, J. W., Bostrom, A., Miles, S., Olsen, M., Gurley, K., Irish, J., Lowes, L., Tanner, T., Dafni, J., Grilliot, M., Lyda, A., and Peltier, J. (2020). Research needs, challenges, and strategic approaches for natural hazards and disaster reconnaissance. *Frontiers in Built Environment*, 6, 1–17. DOI: 10.3389/fbuil.2020.573068.

Williams, P. (1978). Laboratory Development of a Predictive Relationship for Washover Volume on Barrier Island Coastlines. PhD Thesis, University of Delaware, Newark, Delaware.

Wise, R. A., Smith, J., and Larson, M. (1996). SBEACH: Numerical model for simulating storm-induced beach change; Report 4 - Cross-shore transport under random waves and model validation with SUPERTANK and field data. Technical Report, US Army Corps of Engineers, Coastal Research Program.

Wolinsky, M. A. and Murray, A. B. (2009). A unifying framework for shoreline migration: 2. Application to wave-dominated coasts. *Journal of Geophysical Research: Earth Surface*, 114(1), 1–13. DOI: 10.1029/2007JF000856.

Yates, M. L., Guza, R. T., and O'Reilly, W. C. (2009). Equilibrium shoreline response: Observations and modeling. *Journal of Geophysical Research: Oceans*, 114(9), 1–16. DOI: 10.1029/2009JC005359.

Young, R. S., Pilkey, O. H., Bush, D. M., and Thieler, E. R. (1995). A discussion of the generalized model for simulating shoreline change (GENESIS). *Journal of Coastal Research*, 11(3), 875–886.

Zhang, K. and Leatherman, S. (2011). Barrier island population along the U.S. Atlantic and Gulf coasts. *Journal of Coastal Research*, 27(2), 356. DOI: 10.2112/jcoastres-d-10-00126.1.

Zinnert, J. C., Via, S. M., Nettleton, B. P., Tuley, P. A., Moore, L. J., and Stallins, J. A. (2019). Connectivity in coastal systems: Barrier island vegetation influences upland migration in a changing climate. *Global Change Biology*, 25(7), 2419–2430. DOI: 10.1111/gcb.14635.

Chapter 2

Factors in the Formation of Breach Inlets: From Empirical Observations

Timothy W. Kana

Coastal Science & Engineering Inc.
160 Gills Creek Parkway, Columbia, SC 29209, USA

tkana@coastalscience.com

This chapter draws on empirical evidence to illustrate why certain coasts are more vulnerable to breaching. Mixed energy settings are less vulnerable than wave-dominated coasts based on US East Coast examples herein. Coasts with closely spaced inlets are less susceptible to breaching. Barriers backed by marsh-filled lagoons leave no accommodation space for new channels. Breaches tend to heal rapidly under lower tide ranges and inefficient flushing of large lagoons. Inlet stability theory and recent modeling efforts by others show promise for predicting breaches. However, the problem is multifaceted and highly site-specific. Coastal zone managers can track breach vulnerability using simple cross-barrier surveys. Empirical data from New York suggest breaches are unlikely if the subaerial volume exceeds 400 m^3/m. Barrier island disintegration, which is occurring in some sections of the Louisiana coast, is unlikely to occur this century along microtidal high wave coasts such as the Outer Banks of North Carolina because of differences in sediment grain size, longshore transport, and barrier profile geometry.

1. Introduction

Tidal inlets are a key component of barrier island-lagoon systems (Aubrey and Weishar, 1988; Aubrey and Giese, 1993). They provide a hydraulic link between the open ocean and sheltered lagoon waters. As Hayes (1991), FitzGerald (1993), and others have determined, most tidal inlets are formed by two processes: storm-generated scour channels and the closure of estuarine entrances by sand spits. This chapter focuses on breach inlets and their occurrence because they tend to pose a significant but uncertain threat to today's coastal development on barrier islands.

This chapter reviews key factors the author has observed over several decades, and many others have determined empirically or theoretically to place in context why some barrier island coasts are more susceptible to the formation of breach channels. It is not intended to be a thorough review of inlet hydraulics and sediment dynamics but to offer some practical guidance for coastal zone managers charged with safeguarding development and minimizing erosion threats.

Hurricane Hugo in September 1989 impacted the South Carolina (USA) coast and breached a developed spit at the south end of Pawleys Island, destroying several homes and cutting off vehicle access to dozens more (Fig. 1). Yet, as devastating as this event was to a handful of property owners, it remains only one of two such events along South Carolina's developed open coast in the past eighty years. (The other was a breach of a sparsely developed sand spit at the south end of Hunting Island by Hurricane Matthew in October 2016.) Nearly all of the time over the past

Fig. 1.   The breach of Pawleys Island south spit by Hurricane Hugo in September 1989. The breach was closed mechanically about three weeks after the storm (photo courtesy of SCETV).

century, the South Carolina barrier island shoreline has not produced new inlets where development is located.

Breach inlets formed by surges and waves provide obvious and tangible evidence that a particular storm was significant. People often distinguish, or simply remember, a surge event as the one that opened up Inlet "X," just as Hugo did at Pawleys Island. For example, the Great Hurricane of 1938 cut upwards of 10 channels through Westhampton Beach, Long Island, NY, including present-day Shinnecock Inlet (USACE, 1958). Prior to the storm, there was only one inlet — Moriches Inlet — along an 80-kilometer (km) barrier beach extending from Fire Island Inlet to the village of Southampton (Leatherman and Joneja, 1980). That inlet had been formed by a breach of the barrier in 1931 during an extratropical storm. The Great Hurricane of 1938 actually deposited sand in unstabilized Moriches Inlet, choking flows, while numerous breaches formed in the 25 km barrier segment to the east and immediately adjacent to the west (Fig. 2). There were no breach inlets in 1938 along Fire Island, which extends 50 km west of Moriches Inlet. The path of the storm was due north, crossing Fire Island about 15 km west of Moriches Inlet.

As coastal disasters go, a sudden breach channel through a barrier island can be the worst-case scenario for development in its path. However, it can also provide a new way to flush stagnant water out of a lagoon or introduce sediments to the back-barrier area (Hinrichs *et al.*, 2018). Uncertainty about where or when a breach inlet will form tracks with uncertainty regarding the path of storms and associated surge heights. It is broadly accepted that breach inlets are more likely to occur where a barrier island lacks dune protection (i.e., elevation) and where there is tidal surge asymmetry between the lagoon and the ocean (Basco and Shin, 1999).

Management of breach inlets in recent decades has been controversial and generated debate among scientists, engineers, property owners, and government officials. Along most developed barrier beaches, political needs and pressure tend to outweigh environmental issues. Many of today's inlets originated as breaches of barrier islands and then were stabilized by jetties and dredging to accommodate boaters. Florida's (USA) east coast is a prime example. But once inlets are stabilized and development expands around these fixed locations, breaches become problematic.

Storm channels along developed barriers are closed to reconnect roads, reclaim land, and minimize lawsuits (USACE, 1996). Some jurisdictions, such as the state of New York, have established Breach

Fig. 2.   The 25-km barrier island segment that is present-day Westhampton Beach three days after the great hurricane of 1938. About 10 breach channels were opened during the storm, including what has become Shinnecock Inlet (lower right), a stabilized and maintained channel. Moriches Inlet (opened in a 1931 storm) shoaled in 1938 while several channels opened nearby (upper photo). Moriches Inlet was restored and stabilized by jetties in the early 1950s. Images courtesy of USACE-New York District.

Contingency Plans (USACE, 1996; Williams and Foley, 2007; NPS, 2017), which prescribe steps the state or federal government will take to coordinate the closure of a breach or possibly leave it open if it occurs in a wilderness area. Breach management plans are also common around the world to deal with intermittent inlets in settings where open coast waves and littoral transport are large while tidal action is inadequate to maintain permanent channels (Bruun and Gerritsen, 1960; Wainwright and Baldock, 2015; McSweeney *et al.*, 2017).

A motivation for this chapter is the recent evidence for increased storm intensity, accelerated sea-level rise (SLR) (IPCC, 2023), and more frequent barrier erosion events.

Numerous researchers have studied the response of vulnerable coasts where rates of SLR are the highest, such as Louisiana's Chandeleur Islands (Williams *et al.*, 1992; List *et al.*, 1997; FitzGerald *et al.*, 2006, 2007), and documented the disintegration of those barrier islands over the past century (Fig. 3). The general public and many coastal zone managers

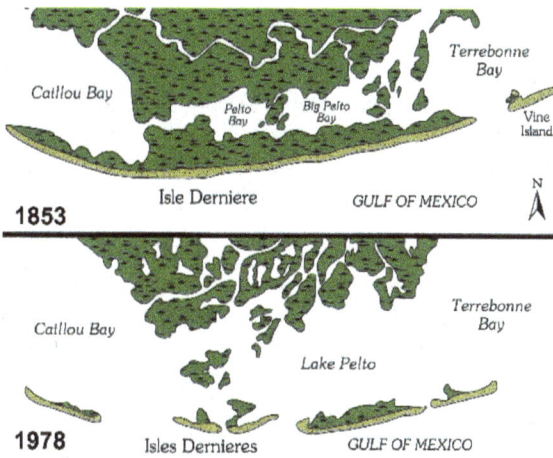

Fig. 3. Shoreline changes along the west end of the Chandeleur Islands, Louisiana, between 1853 and 1978, showing barrier island disintegration and marsh drowning under local SLR, that is upwards of five times greater than global eustatic rates. (After Williams *et al.*, 1992).

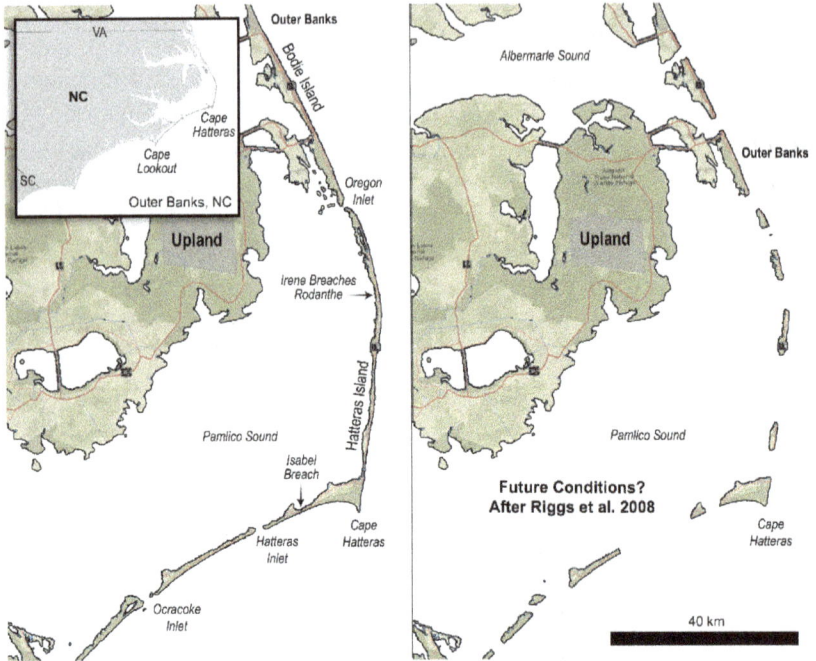

Fig. 4. The Outer Banks of North Carolina (USA) showing existing conditions and hypothesized future conditions under accelerated SLR. This future scenario (after Riggs *et al.*, 2008) speculates that the barrier beaches may disintegrate, leaving remnant islands separated by broad inlets.

can envision parallels between the Louisiana coast and other barrier island shorelines in the future (Fig. 4). However, will that necessarily be the case in the next century or so? Will there be wholesale disintegration of barrier islands under the upper range of likely SLR scenarios (i.e., order of 1 meter/century) (IPCC, 2023)? Evidence from a number of sites, as presented herein, suggests this is likely to be rare, or it can be mitigated through maintenance of the barrier island profile. Most of the examples in this chapter are drawn from Long Island's south shore, the Outer Banks of North Carolina, and the mixed energy coast of South Carolina. The Long Island and North Carolina examples are along microtidal (tide range ~1 meter), wave-dominated (average incident waves >1 meter) settings, while South Carolina is a mixed-energy coast with average tides around 2 meters and incident waves of the order of 0.5 meter.

2. Breach Inlet Distribution

Breach inlets of concern are mostly located along developed barrier islands and spits where private property, public infrastructure, or special back-barrier environments may be adversely impacted. Breach inlets exacerbate problems of erosion by drawing off much more sand from the littoral zone than a typical washover or break through a foredune (Leatherman, 1979). This changes the sediment budget of a barrier beach and locally accelerates erosion rates along the adjacent strand (Liv *et al.*, 1993; Kana, 1995; Rosati *et al.*, 1999, 2013). As breach channels grow, the dominant sediment transport direction is usually into the lagoon, leading to the development and expansion of flood tidal deltas. Fortunately, breach inlets are relatively infrequent along most developed coasts. By far, the greatest occurrences of breaches are along undeveloped, low, narrow spits and barriers.

2.1. *South Carolina: A mixed energy coast*

The mixed energy coast of South Carolina, USA (SC), maintains upwards of 35 tidal inlets along a 300 km ocean coast, of which only four are stabilized by twin jetties and two are armored on the downcoast side to limit migration (Zarillo *et al.*, 1985). Most SC inlets are positionally stable, deep channels anchored in consolidated Pleistocene deposits. As sea levels rose and drowned coastal plain rivers after the last ice age, interfluve areas infilled and formed beach ridge barrier islands (Hayes, 1994). Relatively plentiful sediment supplies produced depositional conditions, so many SC barriers are short and broad, consisting of multiple beach ridges, each representing shorelines over the past 5000 years or so (Moslow, 1980; Hayes, 1994). Tidal prisms for many of the inlets in this mesotidal coast are large (of the order $10^7 - 10^8$ m^3, (Gaudiano and Kana, 2001)), as they drain extensive marsh-filled estuaries under a 2 m tide range. With multiple inlets draining most of the tidal basins, flows are efficient, and phase lags between the ocean and bay tides are relatively low (Nummedal and Humphries, 1978). This lessens the need for breach channels to capture flows and create more efficient pathways for tides exiting the lagoons (Hayes, 1979, 1980).

Most ephemeral, small shallow inlets along the SC coast are located along transgressive barrier beaches where limited sand supplies have produced chronic erosion and left remnant beach ridges plastered against salt

Fig. 5. Edingsville Beach, SC, a transgressive barrier beach dominated by washovers plastered against salt marsh. Breach inlets only occur where the receding coast intersects an existing marsh tidal creek. Otherwise, there is no accommodation space for sediments washed landward to leave a scour channel below low water. (Photo by Coastal Science & Engineering Inc.).

marsh (Fig. 5). Such barriers like Edingsville Beach, illustrated in Fig. 5, exhibit the classic rollover morphology of an erosional shoreface, with washovers perched at wave uprush elevations over marsh deposits which are at mean high tide levels. As the system retreats, buried marsh deposits are re-exposed in the surf zone. Settings such as Edingsville Beach only form breach channels where the retreating barrier beach intersects with a marsh tidal creek. Otherwise, there is no accommodation space for sand scoured from the oceanfront to deposit in the lagoon and create a low tide channel.

Kana *et al.* (2013) reported that roughly 80% of South Carolina's undeveloped shoreline has been eroding in recent decades, whereas a similar percentage of developed shoreline has been accreting, consists of multiple beach ridges, or is maintained by nourishment. This accounts for the exceedingly low incidence of breach inlets of concern in that state.

There appears to be only one segment of the SC coast that fits the barrier disintegration model illustrated in Figs. 3 and 4. Cape Romain, between Charleston and Myrtle Beach, is the southernmost of the four Carolina capes (Cape Fear, Cape Lookout, and Cape Hatteras in North Carolina are the other three). The North Carolina (NC) capes generally accumulate sand under the bi-directional (NE-SW) wind and wave regime of the US East Coast. This convergence of sediment has produced major

Fig. 6. Cape Romain, SC, showing barrier beach disintegration and multiple breach channels through the narrow spits. This area differs from Edingsville Beach (Fig. 5) because there is open lagoon area behind the beach to accommodate overwash and develop viable channels. There is no new sediment coming into the system and net transport diverges away from the apex of the Cape foreland.

offshore shoals at each NC cape. Unlike the major capes, Cape Romain has been sediment-starved over the past century. It maintains an arrow-shaped foreland disconnected from the adjacent strand and is not receiving sand as the other capes are. Instead, sediment transport diverges away from the apex of the cape and builds narrow spits in either direction. Both the limbs of Cape Romain have been breached numerous times and in multiple localities over the past 80 years based on aerial photography (Fig. 6). The breaches have occurred over much of the length of the cape and persist without infill because there is open water rather than contiguous salt marsh behind the barrier spits and a depleted sand supply offshore (Ruby, 1981; Sexton and Hayes, 1996). This provides accommodation space for breach deposits and inhibits recovery or reconnection of the subaerial beach. South Carolina's moderately low incident wave energy retards bar building, and mesotidal conditions maintain strong tidal flows, which inhibit the natural closure of breach channels at Cape Romain.

2.2. *North Carolina: A wave dominated coast*

North Carolina's coastal morphology is quite different than South Carolina's. Mean tide range is roughly half and incident wave heights are double that of SC. The majority of inlets occur along the southern half of the coast (15 over 225 km), where the mean tide range reduces from 1.5 to 1.0 m, and the coast is sheltered from predominant wave energy out of the northeast. Lagoons are generally narrow and marsh-filled south of Cape Lookout. Despite the direct impacts of numerous hurricanes tracking north into the NC coast over the past century, there have been only a few breach channels opened south of Cape Lookout and none along developed barrier beaches. This appears to be due to relatively closely spaced inlets (such as SC) and narrow, marsh-filled lagoons. Even low transgressive barrier beaches such as Masonboro Island have not been breached in the past 80 years (Doughty *et al.*, 2006).

The northern 275 km of North Carolina north of Cape Lookout experiences some of the highest wave energy along the US East Coast (Leffler *et al.*, 1996) and generally consists of narrow barrier islands backed by broad open lagoons. This wave-dominated coast, referred to as the Outer Banks, is relatively sand-rich (Dolan *et al.*, 2016) and maintains only four long-term inlets. Moslow and Herron (1984) estimated that 25% of the Outer Banks barrier lengths consist of inlet fill where breaches once occurred over the past millennium. Mallinson *et al.* (2008) identified 40 sites of inlets that have occurred in the Outer Banks in historical times, with some lasting over a century and others just years before closing. The present four that persist have received maintenance dredging over the past 80 years. Under SLR conditions of the past century (~0.3 m rise), existing inlets and breaches have not shown any tendency to enlarge on their own and capture a significant tidal prism. This is confirmed by the low mean tide range in the sounds, which are of the order 0.15 m, only increasing to ~0.3 m at stations close to the inlets (*Source*: NOAA-tidesandcurrents. noaa.gov/tide_predictions). Nearly all breaches in recent decades have been infilled to restore access roads or closed naturally soon after opening.

The barrier islands at the south end of the Outer Banks between Cape Lookout and Ocracoke Inlet are transgressive and subject to frequent overwash, whereas the northern barriers (Hatteras Island and Bodie Island) tend to have high dunes and greater subaerial volume. Those islands have continuous road access and established communities at the broader sections of each island. Hatteras Island became a US National Seashore in 1953.

Everts *et al.* (1983) reported that the northern Outer Banks have been positionally stable but subject to in-place drowning over the past 80 years. Erosion has occurred along the oceanfront and sound shoreline at low rates (<1 m/yr) on average. Along Hatteras Island, there are erosion hot spots, including Rodanthe (Fig. 4) which have receded at well over 5 m/yr while nearby areas have accreted by almost as much. In short, these largely single ridge barriers have a relatively balanced littoral sediment budget despite high net longshore transport rates in some areas (Inman and Dolan, 1989; NCDCM, 2011; Kaczkowski and Kana, 2012). Erosion-accretion waves (rhythmic topography) propagate alongshore under high wave energy and a plentiful sand supply, healing washovers and the occasional breach inlet before it can widen.

The Rodanthe and "Pea Island" area immediately to the north is notable for multiple breaches in the past century, as documented by Everts *et al.* (1983), Dolan and Lins (1986), Inman and Dolan (1989), Moslow and Herron (1994), Overton and Fisher (2005), and Riggs and Ames (2003). Each event was closed artificially to restore road access along the developed island. One breach in 2011 at a location that had previously been "New Inlet" (Mallinson *et al.*, 2008) was left alone and a bridge built over it (Fig. 7). The breach only lasted two years before infilling naturally, but the bridge remains in place over the washover.

The 2011 "New Inlet" breach (Fig. 7) epitomizes the conflicts in North Carolina regarding breach management. For decades, any breach along 85 km Hatteras Island was closed artificially to restore vehicle access and relink communities that had settled on the island for hundreds of years (NPS, 2015). Some advisors to the agencies that control the National Seashore and a wildlife refuge north of Rodanthe have appeared to recommend that barrier island "rollover" is the natural process which will maintain the integrity of Hatteras Island (NPS, 2020). The apparent

Fig. 7. (left) New River bridge over the Hurricane Irene (2011) breach 8 km north of Rodanthe, NC. Note high wave conditions and shallow depths at the mouth of the channel, which are a prerequisite for bar building and natural closure. (right) Infilled channel and washover under the bridge in 2019. (Photos by CSE and Carolina Designs).

assumption is that the northern Outer Banks are not positionally stable and, like islands to the south, should be allowed to recede via overwash processes (Pilkey *et al.*, 1998; Riggs *et al.*, 2008). Thus, efforts to restore protective dunes or modify the sediment budget in any way are counter to the natural processes molding and shaping that coast. Counter to that argument is the fact that not all barrier islands are transgressive at century time scales as Hayes (1979), Davis (1994), and others have demonstrated. Barrier islands can build seaward via periodic influxes of sand whether by natural or artificial processes as long as the rate of sediment supply exceeds the local rate of SLR (Bruun, 1962).

In this author's experience, much of the conventional wisdom about barrier island rollover has been shaped by the barriers at the south end of the Outer Banks, Core Banks, and Portsmouth Island, which extend 65 km from Cape Lookout to Ocracoke Inlet (Fig. 4). Dunes are poorly developed and each island is dominated by washovers activated almost yearly under storm tides. Typical barrier width above high tide is in the range of 150–200 m, and each island is receding at 1–2 m/yr (NCDCM, 2011). Back barrier areas are predominantly narrow fringing marsh and open water lagoons. The marsh is formed on relict flood tidal deltas (Shawler *et al.*, 2021).

Core Banks became divided by Drum Inlet, a channel that broke through the barrier in 1899, closed naturally in 1919, and then reopened in 1933 during a major hurricane. Since then, the inlet has shifted within a 6 km corridor via artificial dredging (1971) and natural breach processes. Drum Inlet locations over the past century have produced a flood tidal delta spanning upwards of 1,000 hectares (hc) (Fig. 8). Assuming the sand deposits associated with the inlet average a conservative 0.5 m thick, upwards of 250,000 m^3 have been drawn off the adjacent beaches. The small ebb-tidal delta of the inlet has bypassed sand episodically to the downcoast beach (*Source*: earthengine.google.com/timelapse; accessed February 2025). Net longshore transport has generally shifted the channel south in the past 40 years.

Drum Inlet's history and its large flood tidal delta appear to follow a classical breach model whereby a storm breaks through the barrier and a channel is cut via incoming or outgoing tidal surge (discussed in the following section). The breach draws off sand from the adjacent strand, adversely impacting the littoral sand budget more than washovers across the barrier (Leatherman, 1979).

Fig. 8. Drum Inlet, North Carolina, which breached Core Banks in 1899 and has opened and closed naturally at decadal scale intervals. Since the 1970s, it has been maintained sporadically by dredging and held within a 6 km corridor. Note the extensive flood tidal delta and net southerly deflection of the channel under predominant waves from the NE. (Imagery date and source: Google Earth™ 30 October 2017).

While washovers dominate along Core Banks, one recent event has produced over 20 small breach channels across the north end of Core Banks/Portsmouth Island (Sherwood *et al.*, 2023). Hurricane Dorian 5–6 September 2019 tracked NE right over Core Banks. On the approach, winds were directed onshore which activated washovers and lowered water levels on the sound side of the barrier beach. As the hurricane center reached Cape Hatteras, winds in Pamlico Sound and Core Sound to the south backed to the northwest and piled up water against the back side of the barrier. This produced closely spaced breaches directed oceanward at numerous low spots across the island. Imagery from five days after hurricane landfall shows low tide bars at the mouth of each channel (Fig. 9(b)). Two months later (Fig. 9(c)), the bars had coalesced and were deflecting downcoast (south) or closing each hurricane channel. The next available imagery from October 2021 shows all hurricane channels closed

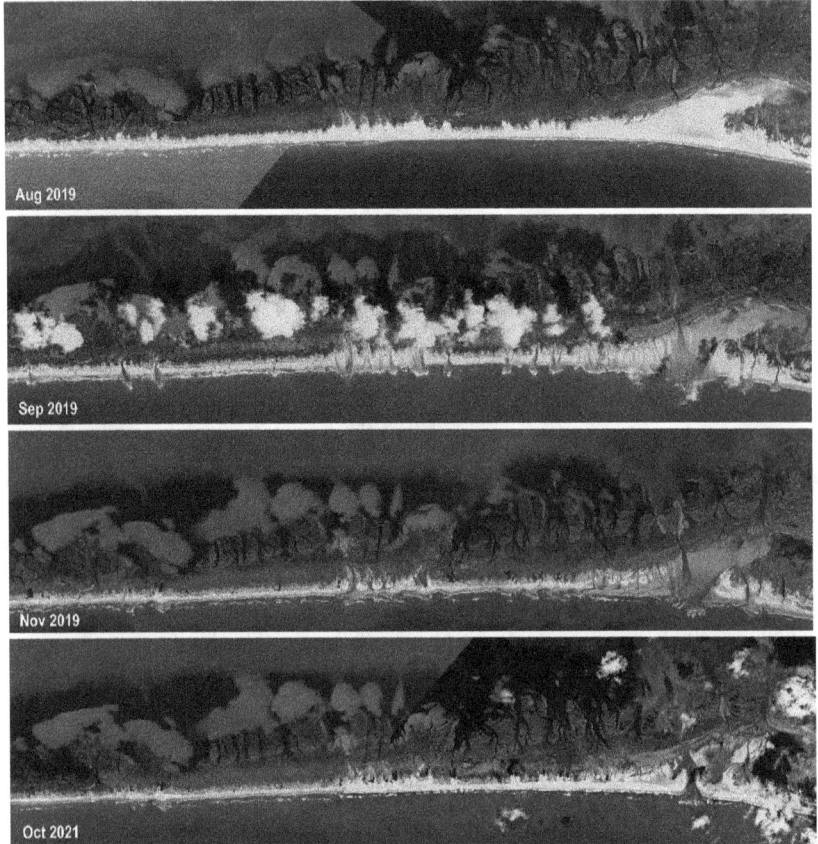

Fig. 9.   Sequence of aerial images before and after hurricane Dorian (5–6 September 2019) at the north end of Core Banks-Portsmouth Island near Ocracoke Inlet. (top) Pre-storm washover barrier, (upper middle) multiple breach channels formed by release of lagoon surge — note ocean shoal development, (lower middle) post-storm coalescing shoals prior to breach closures, and (bottom) infilled channels and return to pre-storm conditions two years after Dorian. (Imagery source: Google Earth™).

and the area returned to its pre-storm condition (Fig. 9(a) and 9(d)). Unlike the evolution of Drum Inlet and its large flood tidal delta, the Hurricane Dorian breaches resulted in minor sand losses to the lagoon. Most of the eroded sand remained in the active surf zone where it was readily available to move onshore and close each breach channel.

Several factors appear to explain why Portsmouth Island had dozens of breach channels formed during Dorian (September 2019), while Core Banks to the south had none. Both washover barriers are similar in width and elevation. In addition to the storm track paralleling the coast and producing a 180-degree wind shift as the eye passed, the area where breaches occurred is exposed to a 40 km (+) fetch across Pamlico Sound. The fetch across Core Sound is only 3–5 km. Shallow lagoons with long fetches experience higher "wind tides" and water level setup. In this case, the surge in the Sound near Ocracoke Inlet was greater than the ocean surge. Sherwood *et al.* (2023) provide a detailed analysis of the Portsmouth Island breaches during Dorian.

The following section discusses the basic processes of inlet breaching and draws on the South Carolina and North Carolina examples to place in context other breach areas.

3. Breach Inlet Processes

It is generally accepted that new inlets of significance are formed where there are existing conditions that allow storm tides to erode protective dunes or first shift sand landward into lower-lying areas of a barrier beach. But to create a new inlet, as Basco and Shin (1999) define, the profile section across the barrier must end up below mean lower low water (MLLW) elevation at the end of a storm. In other words, there should be tidal flows during the entire tidal cycle to be considered an incipient breach inlet. Breaches (including washovers) to some intertidal elevation are likely to shoal and close immediately through influxes of littoral sediment into the low areas. Hayes (1967) noted that full breach channels associated with hurricane Carla (1961) in Texas were initiated by erosion and washovers of the outer beach but cut to below sea level by ebb flows from the lagoon. Asymmetry of the bay and ocean tide/surge phases is a critical factor in inlet initiation since there has to be a hydraulic head in either direction across the barrier to create a channel.

Basco and Shin (1999) broke down the breach inlet formation problem into four stages for a 1-D-coupled numerical model that simulates the
process:

(1) Storm surge and wave attack breaks down the protective dune or
    berm.
(2) Wave runup produces overwash and landward sediment transport to
    lower areas of the barrier island.
(3) Tidal flooding from ocean to bay scours the profile until the water
    levels are equal.
(4) Tidal flooding resumes from bay to ocean as impounded flood waters
    are released through the new gap.

An example result of their simulations is illustrated in Fig. 10, which
tracks the barrier cross-section changes and movement of the volume
centroid over the course of a breach event. A modeled surge hydrograph
for a Basco–Shin simulation is shown in Fig. 11. The elegance of this
early effort to model a barrier breach is that it assumes realistic dimensions for the barrier profile and phase lag between the ocean and bay tides
and the duration of the storm. The initial subaerial barrier cross-section

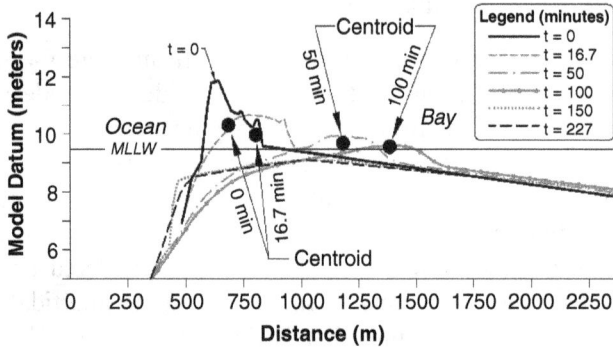

Fig. 10.   Example result of the Basco–Shin coupled 1-D model of a barrier breach,
showing the barrier cross-section evolution (movement of volume centroid above MLLW)
over an approximately 4-hour simulation. The breach is initiated as erosion and loss of the
foredune, followed by landward translation of material into the bay (lagoon), then final
scour of the subaerial profile to below low tide. Other simulations by Basco and Shin
demonstrate the importance of the bay and ocean phase lags and duration of the storm to
effect a full breach. (After Basco and Shin, 1999).

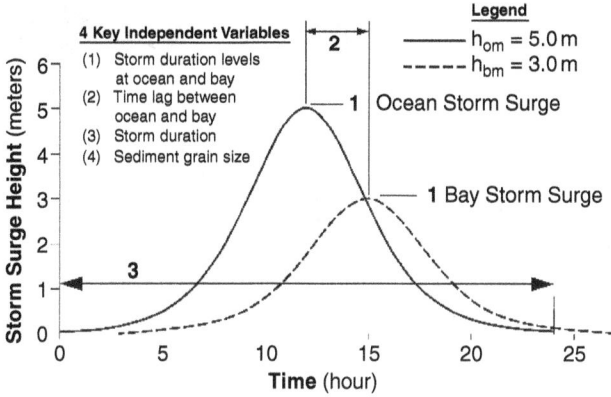

Fig. 11. Characteristic ocean and bay (lagoon) surge hydrographs used in simulations like the results in Fig. 10. The phase lag between ocean and bay tide creates the hydraulic gradient (head of water) necessary to cut breach channels from the ocean and or bay side. (After Basco and Shin, 1999).

(above MLLW) is ~400 m^3/m. The surge heights in the ocean and bay and phase lags were modeled for ranges of 3–5 meters and 1–3 hours, respectively, which is realistic for many US East Coast sites. The key findings were the importance of the peak surge difference from ocean to bay, the phase lag between the ocean and bay surge heights, and sediment grain size. Small phase lags (of the order of one hour), which result more often where multiple inlets occur to handle the local tidal prism, or where the entire barrier is overtopped by a surge, were less likely to generate a complete breach compared with larger phase lags (order of three hours) for the same surge heights (Basco and Shin, 1999). This makes perfect sense and is consistent with the observation of channel cutting during the "back" of the storm when impounded lagoon waters are channeled through breaks back to the already receded ocean.

Hurricane Isabel, 18 September 2003, breached the south end of Hatteras Island (Fig. 4) as a Category 2 storm with maximum winds around 165 km/hr. The peak surge was 1.5 m above predicted tide and in phase with the time of high tide (Walmsley and Hathaway, 2004). The multiple-channel breach occurred at a low, narrow section of the island that had breached in 1933 (Fig. 12). The island at that locality was only 150 m wide and lacked dune protection partly because the oceanfront area includes a large parking lot and pathways for the convenience of

Fig. 12. Aerial photo dated 21 September 2003 of Hurricane Isabel breaches at south Hatteras Island. *Source*: US Geological Survey.

beachgoers. Strong onshore flows initiated the breach and it expanded quickly into open sound waters. The breach was closed by dredge 1.5 months after the event. It appears the breach channels were viable conduits before the ocean tide receded. Subsequent ebb and flood flows served to enlarge the channels and initiated the formation of flood and ebb tidal deltas (Walmsley and Hathaway, 2004). Upwards of 340,000 m^3 were dredged from the flood shoals to fill the breach channels (Wutkowski, 2004).

The storm-surge hydrograph of Fig. 11 is fairly representative for many US East Coast barrier island sites. However, even the lagoon surge is likely to be more asymmetric and flatter. The peak ocean surge will develop rapidly and usually exceed the bay surge, which is dependent on the rate of flow through existing inlets that are often many kilometers (km) apart. As hurricanes move out of an area, the ocean surge recedes rapidly while the lagoon empties more slowly, prolonging the duration of the ebb discharge (Nummedal and Humphries, 1978).

Hurricane Irene Breaches

While Basco and Shin (1999) used realistic values for their surge hydrographs, empirical evidence from some sites shows breaches can occur under lesser storms with lower peak surges. In some settings such as the Outer Banks or Padre Island, Texas (Hayes, 1967), a storm does not have to generate an extreme ocean surge to produce a breach. The "wind tides" in the lagoons, as previously mentioned, can produce a head of water

Fig. 13. Hurricane tracks over North Carolina for Isabel (2003) and Irene (2011), both of which produced breach inlets through Hatteras Island. Aerial of the Irene breaches at north Hatteras Island on 2 September 2011. (Photo by CSE – H. Kaczkowski).

between the bay and the ocean, driving flows across low areas of the barrier. This appears to have been an important factor in the breaches at north Hatteras Island during Hurricane Irene (2011) (McNinch *et al*., 2012) (Fig. 13). As the storm tracked north (Fig. 13), winds shifted to westerly components and raised the tide along the backside of Hatteras Island north of Cape Hatteras. This produced a prolonged head of water from the sound to the ocean. Breaches occurred where the foredune had been cut away, leaving an easy path for the ebb flows to form new channels.

Irene's storm track (Fig. 13) caused a major water level drop in the lagoon on the east side of Pamlico Sound as the storm approached. While the ocean surge (~2.0 m MSL) from this Category 1 hurricane (at landfall) was well below the predicted 100-year still water surge level, the low water in the sound enhanced the head from ocean to sound. Once the storm passed to the north, winds veered to the west and produced a dramatic wind tide or seiche against the backside of the barrier. In Duck NC, approximately 60 km north of the Irene breaches, tides rose from 0.5 m below MSL to nearly 2.0 m above MSL in less than four hours (McNinch *et al*., 2012). Close to Oregon Inlet in Manteo Harbor, there was an even higher water level differential between the lowest and highest tide levels as the storm tracked north, according to CSE field personnel who rode out the storm on board a vessel (P. McKee, pers. comm. September 2011) and monitored a tide staff. This lagoon flooding lapped over the back barrier flats and produced oceanward flows through dune breaks. At least three of these breaks became breach channels (as defined by Basco and Shin (1999)) within hours of the storm passage (Fig. 13).

Cupsogue NY Breach Case

Another breach example that does not fit conventional theories of formation occurred in 1980 near Moriches Inlet, New York (USACE, 1980; Vogel and Kana, 1985). Moriches Inlet formed as a breach of the barrier beach in 1931. Local interests sought to keep the inlet open via dredging and then the construction of parallel jetties in the early 1950s. When the inlet formed, there was a major deposition of sand in the lagoon (Moriches Bay) and the formation of a flood tidal delta. One characteristic of these bay shoals is the development of flanking channels around the horseshoe-shaped shoals to carry flows around the delta (Hayes, 1975, 1980). The eastern bay channel of the inlet encroached on the back side of the upcoast fillet of the jetty in the area known as Cupsogue Spit. While the fillet impounded some sand on the ocean side of the spit and advanced the shoreline, the bay shoreline receded (Fig. 14). In January 1980, the narrow point in the spit was breached during a minor storm. The channel proceeded to widen in the next year and ultimately merged with nearby Moriches Inlet. In this case, there could not have been a significant head difference between tides through Moriches Inlet and flows through the incipient

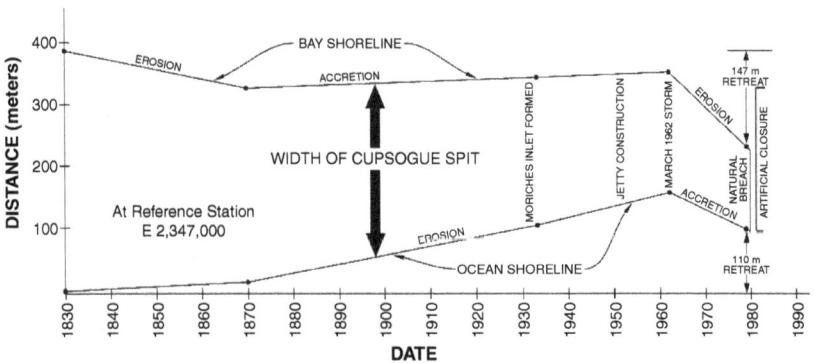

Fig. 14.   Time history of Cupsogue spit, the east fillet of Moriches Inlet NY from 1930 to 1980 when the spit breached. At the time of the breach, the spit was less than 30 m wide at the breach point. (After Vogel and Kana, 1985).

breach because they were so close together. However, a key factor in the development of that breach was the proximity of a relatively deep bay channel that could receive sands washed through the gap. Vogel and Kana (1985) estimated the breach deposited over 575,000 m³ into Moriches Bay during and after the storm before closure was attempted. The sequence of aerials in Fig. 15 illustrates the conditions before the breach was closed by dredge, and the backshore of the spit was armored. Shoals across the breach show that the shallow platform had not captured all the tidal prism of Moriches Inlet but likely exacerbated shoaling of the maintained inlet.

3.1. *Inlet stability*

Thanks to the work of O'Brien (1931, 1969), Escoffier (1940, 1972), Bruun and Gerritsen (1960), Keulegan (1951, 1967), O'Brien and Dean (1972), Jarrett (1976), and others, the scale and stability of inlets is fairly well known. A breach channel necessarily begins with a tiny cross-sectional area, and to become a viable channel, it scours and enlarges, carrying an increasing tidal prism. Escoffier (1940), Keulegan (1967), and O'Brien and Dean (1972) developed empirical relationships to predict the tendency for inlet channels to scour, shoal, or become essentially self-maintaining. They determined that an inlet will have some maximum critical flow velocity (Vmax), which is a function of tidal amplitude differences between the ocean and lagoon (i.e., the primary hydraulic force to drive flows), the inlet cross-sectional flow area, the hydraulic radius of the inlet channel, and the length of the channel. These latter variables add friction factors that dampen flows through the channel. Shallow, long channels provide more resistance to flow than short, deep channels (see source references).

Figure 16 illustrates an example of "stability curve" for an inlet as derived from the empirical equations of Keulegan (1967) and O'Brien and Dean (1972). Until the critical maximum flow velocity and flow cross-section are reached, a new tidal channel will be unstable either in the scour mode or the deposition mode on the left side of the curve (Fig. 16). In the scour mode, the new channel enlarges, and Vmax increases since there is a tendency for decreasing friction as the flow cross-section also increases (Fig. 16). If the cross-sectional area reaches or exceeds a critical flow area (Ac) for the setting (function of the local tidal prism), further enlargement of the channel (right side of the curve) will lead to shoaling. That is,

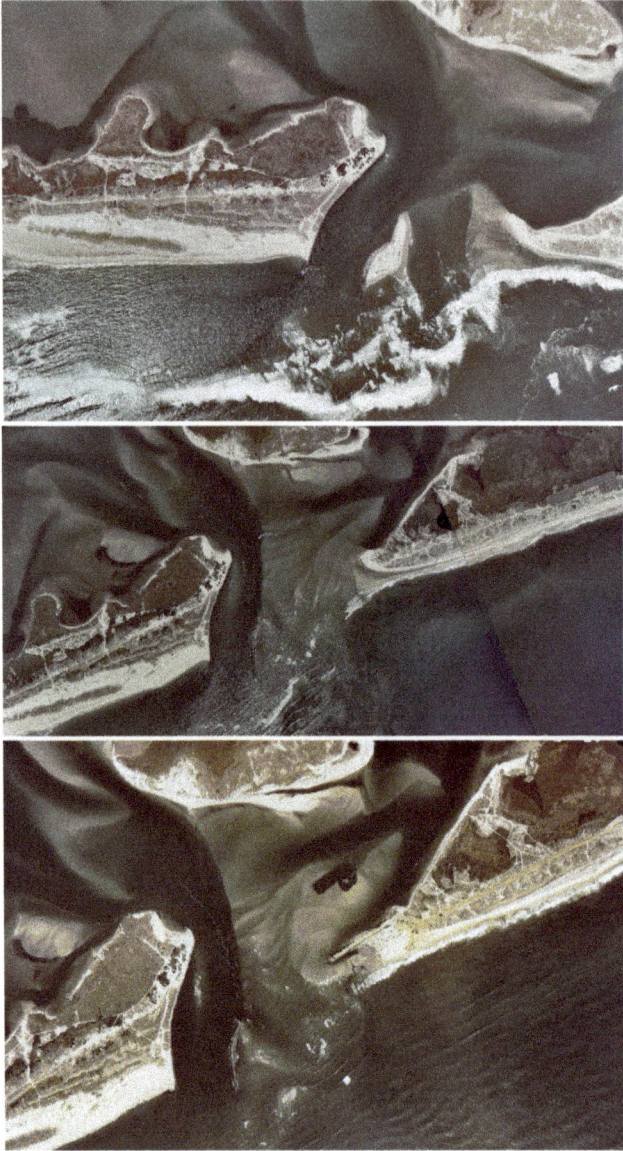

Fig. 15.   The Cupsogue breach in the east fillet of Moriches Inlet in 1980: (top) 14 January just after the opening; (middle) 18 February after channel widening to the east jetty; (bottom) 20 October at the beginning of closure operations by dredge using shoals of the flood delta that were deposited by the breach. (Imagery courtesy of USACE, NY District).

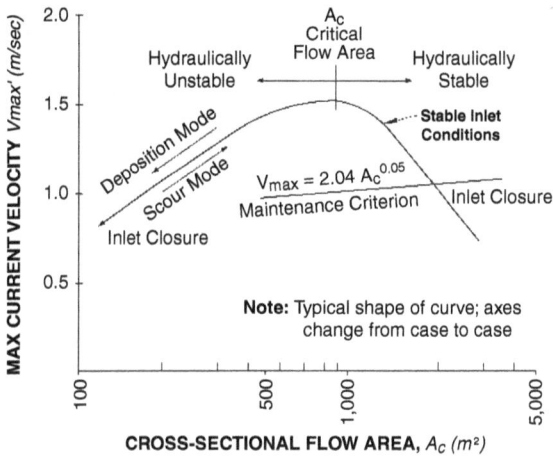

Fig. 16. Inlet stability curve after O'Brien and Dean (1972) illustrating the relationship of flow cross-section to velocity. The left side of the curve represents hydraulically unstable conditions of scour as the channel enlarges or deposition as velocities decrease. The right side of the curve represents the stable inlet condition whereby the channel section is self-maintaining in response to variations in current velocity.

velocities decrease because of the larger channel cross-section, beyond that which is necessary to handle the volume of water ebbing and flowing with each tide. But shoaling and reduction in the cross-sectional area on the right side of the curve lead to scour, so the channel adjusts to velocity changes in a hydraulically stable manner. The majority of South Carolina inlets are classed as hydraulically stable and self-maintaining (Nummedal and Humphries, 1978) because the ~2 m tide range and extensive marsh-filled lagoons lead to relatively high Vmax, well above the critical velocity for sediment transport.

Czerniak (1976, 1977) applied the O'Brien and Dean (1972) stability curve to the case of Moriches Inlet. That inlet has been difficult to maintain because of its tendency to shoal and close. Between 1931, when it was first breached, and 1938, it was hydraulically unstable in the scour mode. As the channel scoured, the tide range in Moriches Bay increased, leading to a larger tidal prism and higher velocity. The 1938 hurricane opened other short-lived inlets nearby (Fig. 2), captured some of the tidal prism of Moriches Bay, and led to shoaling in Moriches Inlet during the 1940s. The shoaling reduced flows and lowered the tidal prism passing through the channel. The bay tide range decreased, and the channel

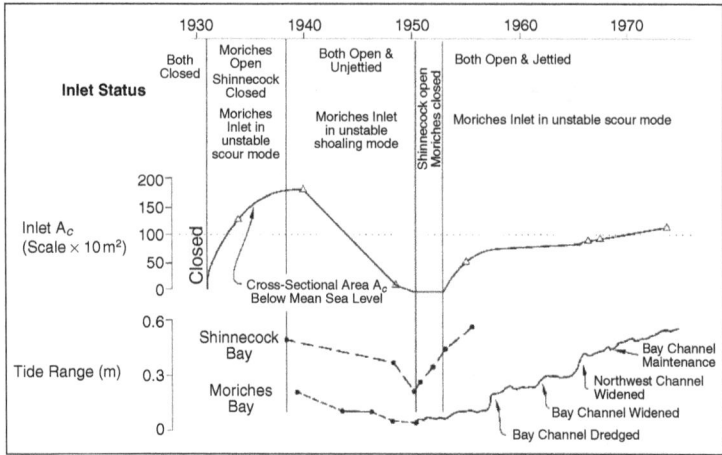

Fig. 17. Moriches Inlet (NY) history and related changes in flow cross-section (Ac), bay tide range, and hydraulic stability. Increasing flow efficiency via dredging led to higher bay tidal range and revision of the stability curves (Fig. 18) (based on (Czerniak, 1977)).

closed. This led to the construction of jetties and a program of dredging, which continues to the present. Channel dredging has led to increased bay tide range and decreasing frictional resistance (Fig. 17). Czerniak (1977) showed how these changes modified the stability curves and increased Vmax (Fig. 18). His results demonstrate how inlet stability is a "moving target" and will depend highly on the tide range in the bay.

In the author's observation, low tide range settings have prolonged slack water periods over the fortnightly tidal cycle, which increases the time over which waves and littoral sand transport can infill unstable channels. Long channels in microtidal settings similarly tend to experience greater frictional effects on flows, leading to more rapid shoaling and closure than higher tide range areas. In mesotidal mixed energy settings such as South Carolina, the duration of slack water periods (velocities below the critical shear stress for sediment transport) will be shorter. Introduced sediments during these slack periods are more likely to be scoured away, and the channel is maintained during the high-velocity portions of the tidal cycle. Some small inlets in South Carolina, such as Pawleys Inlet (Ac <100 m² – (Gaudiano and Kana, 2001)), are migratory but persist under daily tides. A similar scale channel in a microtidal setting, where lagoon tide range is small, is more likely to shoal and close.

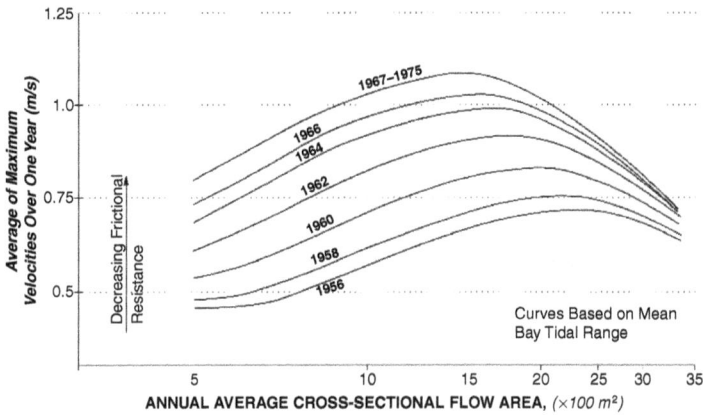

Fig. 18. Moriches Inlet stability curves for the period of 1956–1975 as reported by Czerniak (1977) and converted to metric units herein.

4. Identifying Potential Breach Points

It is well established that washovers and breaches are relatively common and frequent along transgressive barrier islands but rare along regressive (i.e., prograding) coasts (Pierce, 1970; Hayes, 1991; Donnelly *et al.*, 2006; Lorenzo-Trueba and Ashton, 2014; Nienhuis *et al.*, 2021). Coasts with a positive sediment budget or ongoing artificial nourishment are likely to have an area of dry sand beach during the entire tidal cycle. Aeolian transport across the dry beach will feed dune growth and increase resilience during storm surges.

Bocamazo *et al.* (2011) documented dune growth of ~3 m³/m during the first four years of a nourishment project at Westhampton Beach (NY), the site of prior breach inlets. Kaczkowski *et al.* (2018) measured even higher rates of dune growth following a 2011 nourishment project at Nags Head (NC) and correlated rates with berm width (i.e., "fetch") using Bagnold's (1941) formulation. If dunes grow and overtopping diminishes, littoral volumes tend to be preserved seaward of the dune crest, and net annual erosion rates are lessened (Kana, 1995). However, once a barrier island loses dune protection and goes into "washover mode," breaches will be more likely.

In the case of Long Island, New York, the long continuous segment of barrier beach between Fire Island Inlet and Southampton (Fig. 19) did not

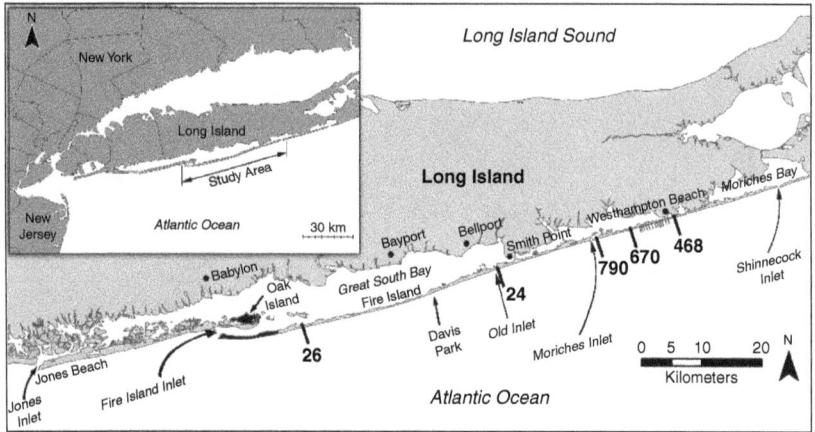

Fig. 19. Location map of representative barrier-island cross-sections referenced herein. The barrier beach from Fire Island Inlet to Southampton (at right edge of map) was continuous between 1825 and 1931 when Moriches Inlet formed.

breach from around 1825, when an inlet off Bellport closed, to 1931, when Moriches Inlet opened (Leatherman and Joneja, 1980). Immediately west of Fire Island Inlet, there was a low barrier island with intermittent channels in the area now known as Oak Island to Jones Beach. This shoreline was reportedly subject to frequent washovers in the 1800s (Moses, 1939; USACE, 1957). In the late 1920s, nearly 30 million (M) cubic meters were dredged from Great South Bay and deposited along and across the entire barrier (Moses, 1939). Like Galveston Beach after the 1900 hurricane, the land was raised upwards of 4 m above mean high water, intermittent channels were filled, and a 20 km long highway was built along the spine of the island. This greatly increased the subaerial volume and led to a century with no breach channels or cross-barrier overwash.

The present Fire Island segment of Long Island's south shore is similarly configured as Hatteras Island, North Carolina, but with a much smaller lagoon. It consists of a relatively narrow barrier (200−400 m wide) backed by 3−9 km-wide Great South Bay. Fire Island experienced the major storms of 1938, 1962, and 1991−1992 but did not breach until Hurricane Sandy in October 2012 (Hapke *et al.*, 2017).

Leatherman and Joneja (1980) confirmed the general stability and low erosion rates along Fire Island. Profiles along much of the beach exhibited the classic form of a microtidal barrier island cross-section with a high dune ridge along the ocean coast and low relief and elevations across the back barrier. Some sections of Fire Island reached elevations >20 m above MSL in the 1960s.

Rosati *et al.* (1999) and Kana *et al.* (2011) attributed the stability of Fire Island to a balance of longshore sand transport bypassing Moriches Inlet and generally moving to the west. Others have attributed its stability to onshore transport from deep water (Schwab *et al.*, 2013). Regardless of the cause, shoreline stability allowed a well-developed foredune to per-sist. In recent years, beach nourishment projects have further enhanced the sand supply and helped maintain the protective dune along developed portions of the island (i.e., from Davis Park to Fire Island Inlet, Fig. 19). However, this has not been the case along some wilderness sections, par-ticularly the area off Bellport around Old Inlet.

Kana and Mohan (1994, 1996) analyzed Fire Island and Westhampton Beach profiles for erosion and breach vulnerability. A goal of the study was to predict where breaches were likely and to what extent a breach would increase flooding on the mainland shore of Great South Bay, the lagoon protected by Fire Island. They used a volumetric approach and barrier island cross-sections from nearby areas that had breached to estab-lish threshold criteria. While this method is site-specific for the particular profile geometries of the area, it can be applied to other settings once suf-ficient survey data are available. Through various federal projects, includ-ing a regional sediment budget for the south shore of Long Island (Kana, 1995), Kana and Mohan (1996) had access to pre-and post-breach data for events at Pikes Beach on Westhampton (1962, 1992) and the previously described Cupsogue breach (1980). Figure 19 shows the setting and loca-tion of certain transects referenced herein.

Westhampton Beach experienced a breach at transect WH670 (Pikes Beach) during the March 1962 "Ash Wednesday" storm (USACE, 1963, 1967). The breach occurred at a low, narrow section of Westhampton Beach prior to the construction of 15 groins in the mid-1960s (Nersesian *et al.*, 1993). After the groins were constructed (without concomitant nourishment), Pikes Beach remained narrow, low, and vulnerable to over-topping (Fig. 20). The 1962 breach was closed mechanically within weeks of its opening before it could significantly widen and deepen. The cost of closure was (~)$300,000 (1962 US Dollars) (USACE, 1980; Sorensen and Schmeltz, 1982). The 1992 breach was in the same location, but there were administrative delays before agreements were in place to close it. Nearly 10 months passed before equipment and dredges were mobilized.

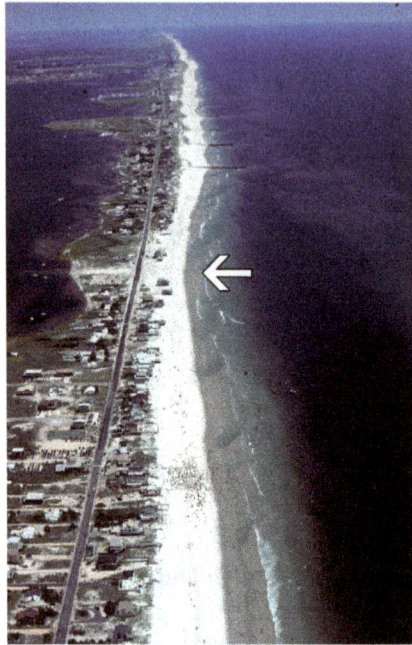

Fig. 20. Low-tide aerial photo in August 1980 of Westhampton Beach (Pikes Beach) at the downcoast end of the groin field (top of image). Breaches occurred in this area in 1938, 1962, and 1991/1992. (Photo by the author).

By then, the breach channel had widened to nearly 600 meters, and a large flood tidal delta had built up in Moriches Bay (Kraus and Wamsley, 2003). About 1.5 years after the 1992 breach, the new inlet was closed at a total cost of (~)$11 million (1993 US Dollars). Much more sand was required to fill the 1992 breach compared to the 1962 breach. The only section of Westhampton Beach that breached in 1992 was the sediment-starved area immediately downcoast of the groin field. This is in stark contrast to the 1938 storm impacts when about 10 breach channels formed along the west half of Westhampton Beach during that event (USACE, 1958) (Fig. 2).

Two transects illustrate the differing vulnerabilities of Westhampton Beach after 1962 (Fig. 21(a)–(b)). Station 468 is a bay-to-ocean transect out to the estimated local depth of closure (approximately −10 m MSL) in 1962 and 1979. The groin field was constructed between 1964 and 1969. Profile volumes spanning certain contour intervals are given in Table 1 for each date. As shown in Figure 21, there was a major accumulation of sand

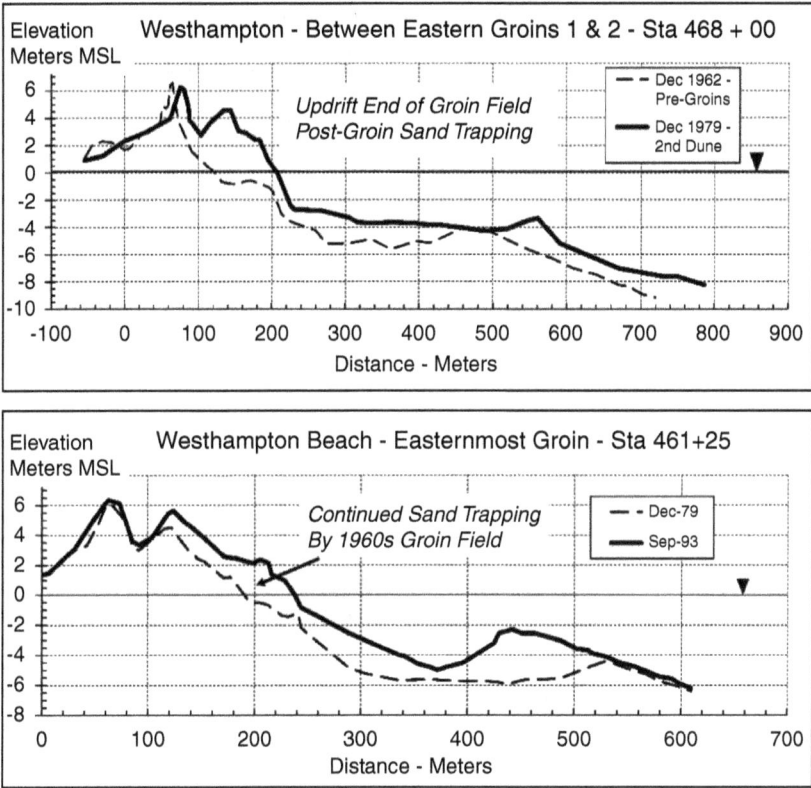

Fig. 21. Profiles from the updrift end of the Westhampton Beach (NY) groin field showing decadal scale accretion and dune growth, increasing the barrier island resilience during storms. (upper) 1962 and 1979; (lower) nearby station 1979 and 1993 (after Kana and Mohan, 1996).

across the littoral profile. In 17 years, the subaerial barrier volume (above MSL) increased by over 50% and a new dune ridge formed 75 m seaward of the 1962 dune crest. A nearby transect around Groin #1 shows continued profile buildup between 1979 and 1993 (Fig. 21(b)). In short, profiles within the groin field gained sand steadily between 1962 and 1993, becoming more resilient to erosion and barrier breaching. Groins #1 and #2 (easternmost) extended 150 m offshore when built but were completely buried by 1994 (Kana and Mohan, 1996).

By comparison, transect 670 at Pikes Beach lost volume between 1962 and 1979 (Fig. 22). The 1962 foredune (to 5 m MSL) was washed

Table 1. Representative barrier island cross-sections. A prerequisite for breaches for this section of the coast is unit volumes below 400 m³/m.

Profile	Date	Above 3 m MSL	0–3 m MSL	Notes
		Unit Volume m³/m		
WESTHAMPTON BEACH				
WH 468 + 00	1962	88.8	416.3	Updrift end of groin field
	1979	130.2	645.0	Updrift end of groin field
WH 461 + 25	1979	119.4	474.1	Updrift end of groin field
	1993	206.7	611.7	Increased dune profile
WH 670 + 00*	1962	35.8	402.2	Pikes Beach downdrift of groin field
	1975	0.0	292.3	Pikes Beach downdrift of groin field
	1979	0.8	394.8	Includes Minor Nourishment
WH 790 + 00**	1974	55.1	407.3	Cupsogue Spit
	1980 (Jan)	52.9	348.8	Pre-breach
	1980 (Jul)	0.0	0.0	Post-breach
FIRE ISLAND				
FI S151 + 024***	1967	91.8	525.0	Old Inlet
	1979	101.8	618.2	Old Inlet
	1993	0.0	393.3	Pre-breach
FI 26	1967	60.0	1,173.5	Saltaire
	1979	55.2	1,177.8	Western Fire Island
	1993	7.8	1,102.5	Broad low backshore

Notes:* Breaches – 1962, 1992 – Major Nourishment and Restoration 1997 – Downcoast of groin field
** Breach – Jan 1980 – Due to Encroachment of Bay Channel – Near Moriches Inlet
*** Breach – Oct 2012 – Hurricane Sandy – Prior opening 1770–1825.

out in minor storms, and a minimal profile (without dune) was maintained in the 1970s and 1980s by small-scale truck fills or scraping to protect the access road. The 1992 nor'easter caused at least two breaches between transect 670 and the westernmost groin (#15). The one closest to the groin became dominant and expanded from about 100 m wide after the storm to over 600 m wide by the summer of 1993 (Terchunian and Merkert, 1995). The 1992 breach inlet continued to widen and deepen, consistent with the

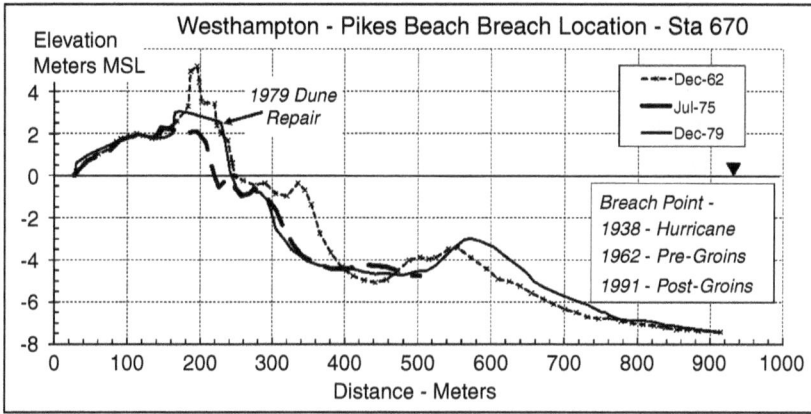

Fig. 22. Profiles from 1962 to 1979 immediately downcoast of the Westhampton Beach groin field showing decadal erosion and loss of dune protection, leaving the subaerial barrier volumes below a resilient level (Table 1).

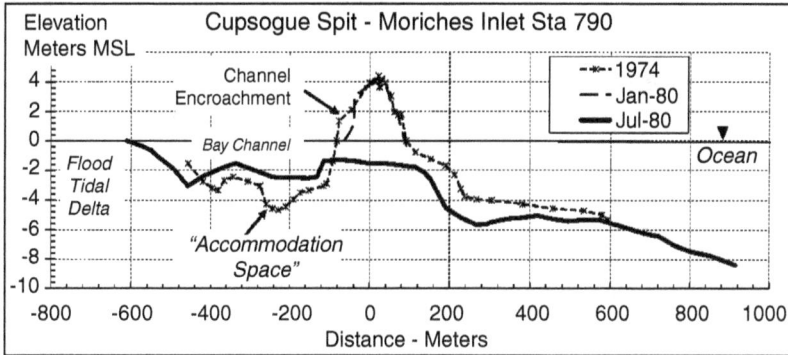

Fig. 23. Profiles adjacent to the 1980 Cupsogue Spit breach. The initial breach point was about 200 m east of this transect, where a broad erosional arc from the bay channel had cut away the back side of the dune ridge. The breach widened quickly and migrated through the section (Fig. 15) (after Kana and Mohan, 1996).

stability theory for an unstable channel (Figs. 16 and 17). Around November 1993, the breach was closed via hydraulic dredge using sand from a federal offshore borrow area (Conley, 1994).

Another set of barrier transects was available for the area near the Cupsogue breach in 1980. Figure 23 shows conditions in 1974, January

1980 (before the breach), and July 9, 1980 (5.8 months after the breach). Section 790 is actually situated between the initial breach point and the east jetty of Moriches Inlet. The breach occurred at the narrowest section of the spit along a broad erosional arc that had formed via encroachment of the eastern bay channel (Fig. 15). The subaerial cross-section at the breach point was a fraction of the profile volume at Station 790. The breach quickly enlarged and passed through this transect (no pre-breach data were available at the actual breach point). Figure 23 shows the post-breach recession of the foreshore and deposition in the bay shoals. The lagoon at that locality, importantly, had accommodation space for washover deposits. Along barrier coasts backed by marsh-filled lagoons (e.g., Edingsville Beach, South Carolina) or shallow flats (e.g., most of Fire Island), the accommodation space for sediments from a breach will be limited. This, in turn, lessens the chance of the breach becoming permanent.

Kana and Mohan (1996) determined that Westhampton Beach, the barrier island that had experienced over a dozen breach inlets during the 20th century, contained significantly less volume in the barrier profile than neighboring Fire Island, which did not breach during the 1900s. Typical profile volumes for Westhampton Beach and Fire Island are given in Table 1 (see (Kana and Mohan, 1994, 1996) for additional details). While the data were limited, site-specific, and had high standard deviation, they provided a simple measure of sand quantities that would be required to bring Westhampton up to the average profile volume along Fire Island (thereby reducing breach vulnerability and increasing resiliency of the barrier island).

Using the data in Table 1, Kana and Mohan (1994) identified two sites–Water Island near Davis Park and "Old Inlet" (Fig. 19) as being most vulnerable to a breach. At each site, the barrier volume above MSL had reduced to levels comparable to low areas along Westhampton Beach (prior to breaches) due to erosion and lack of nourishment.

Figure 24 shows two localities on Fire Island where barrier island resilience decreased in the 1990s prior to Hurricane Sandy (2012). A transect at Old Inlet (USACE Station 151+024) had a healthy foredune through the 1980s. Volumes above the 100-yr return period flood level (~3.0 m MSL) were upward of 100 m^3/m (Table 1). By 1993, the foredune was completely eroded, and the subaerial volume across the barrier was reduced to less than 400 m^3/m. Under a policy of no nourishment within the Fire Island National Seashore, and little natural recovery after the 1991–1992 storms, the sand deficit persisted at Old Inlet. Hurricane

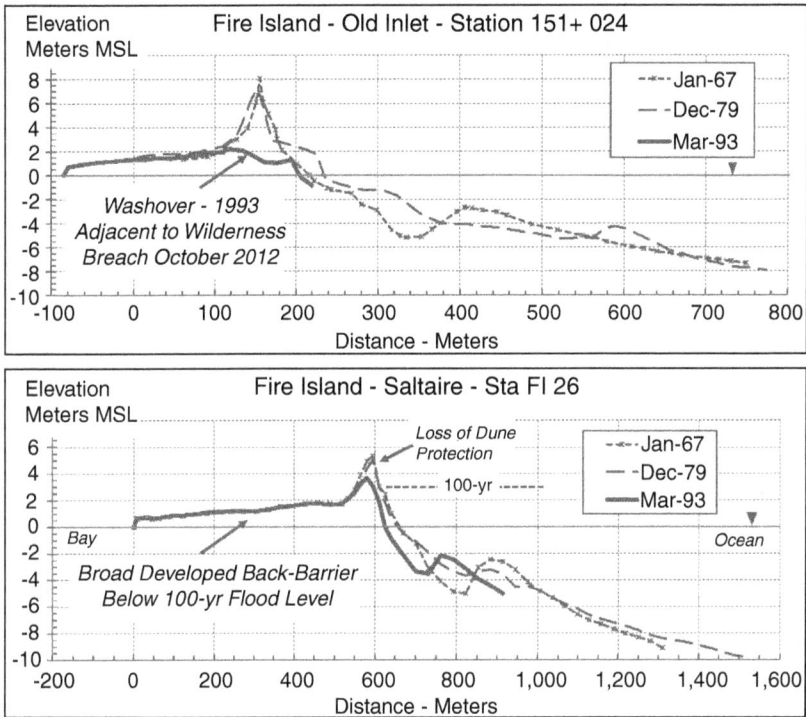

Fig. 24. Representative cross barrier profiles from Fire Island (1967–1993) illustrating loss of dune protection and subaerial volumes above MSL (Table 1). Erosion at Old Inlet (upper) reduced volumes to levels at Westhampton Beach where breaches had occurred. The lower profile shows a locality along western Fire Island where the foredune was severely eroded but subaerial volumes above MSL remained high because of barrier width, reducing the likelihood of a breach (after Kana and Mohan, 1994, 1996).

Sandy produced a surge that easily overtopped the berm at Station 151 + 024. A shallow bay channel to a public dock at Old Inlet provided accommodation space for washover sediments to shift into the lagoon and produce a full breach before the end of the storm.

Another profile (USACE 26, see Fig. 19) across the Village of Saltaire in western Fire Island also lost its dune protection between 1967 and 1993 (Fig. 24, lower). However, that locality did not breach during Sandy because its back barrier was much wider than Old Inlet. Subaerial volumes above 3 m MSL decreased from 60 to 8 m^3/m between 1967 and 1993. However, the cross-barrier volume above MSL remained well above

1000 m³/m, or 2.5 times the unit volumes of vulnerable breach areas at Westhampton Beach, Old Inlet, and Water Island (not shown).

From a coastal zone management perspective, there was considerable debate in the 1990s on whether to maintain the developed beaches along Fire Island via nourishment (USACE, 2020) or allow erosion to progress naturally (Rather, 2001). Fire Island officials with the National Seashore faced the difficult task of balancing the interests of private homeowners, who controlled roughly 40% of the island's length, and the mandate of the federal government to minimize or prohibit erosion control measures within the Seashore. Property owners wanted to maintain the beach and protect their properties, while local Park officials actively pursued a policy of letting nature take its course, even if it meant erosion, barrier breaches, and some property abandonment (USACE, 2020). The 1991–1992 storms raised concerns that a breach of Fire Island would potentially raise storm tide levels around Great South Bay and lead to much greater damages (Cashin Associates, 1993; GCETF, 1993; LIRPB, 1994; Conley 2000; Irish and Canizares, 2009).

Wilderness Breach Evolution

The "Wilderness Breach" reopened Old Inlet on Fire Island nearly 190 years after it closed in 1825 (Hapke *et al.*, 2017). Another near-breach occurred 7 km to the east during Hurricane Sandy, but it did not meet the Basco and Shin (1999) definition of an inlet with some flow at MLLW. The east Fire Island breach immediately closed, whereas the Old Inlet breach off Bellport enlarged and formed a 200-hectare flood tidal delta over the pre-existing "flats" of the back-barrier area (Flagg and Flood, 2013; Van Ormondt *et al.*, 2015).

Hapke *et al.* (2017) noted that the initial breach channel did not occur at the lowest point in the dunes but formed in a location of a cross-island boardwalk that Sandy destroyed.[1] Figure 25(a)–(c) shows the evolution of the breach between November 2012 and 2016. At its maximum, the breach was about 450 m wide and up to 6 m deep in the throat section (Flagg and Flood, 2013). Based on Google Earth™ imagery, the channel expanded for about seven years and then began infilling; by winter 2021, the channel width was halved; by October 2022, the channel was

---

[1]Nummedal *et al.* (1980) reported similar favoritism of washovers and breach channels across Dauphin Island, Alabama, during hurricane Frederic (September 1979). Paved driveways to oceanfront cottages off the main access road turned into efficient flood ramps for the flood and ebb surges.

Fig. 25. The Wilderness Breach at Old Inlet off Bellport, NY, on (top) 11 November 2012, (middle) 27 January 2013, and (bottom) 8 November 2016. (Images: SUNY Story Brook University).

effectively closed; and by June 2024, the entire floodway filled to sub-aerial "dry sand" berm elevations (Fig. 26).

The opening and closing of the Wilderness Breach occurred over a 10-year span. Morphologic evidence suggests the channel originally captured over a half million cubic meters from the adjacent barrier beaches

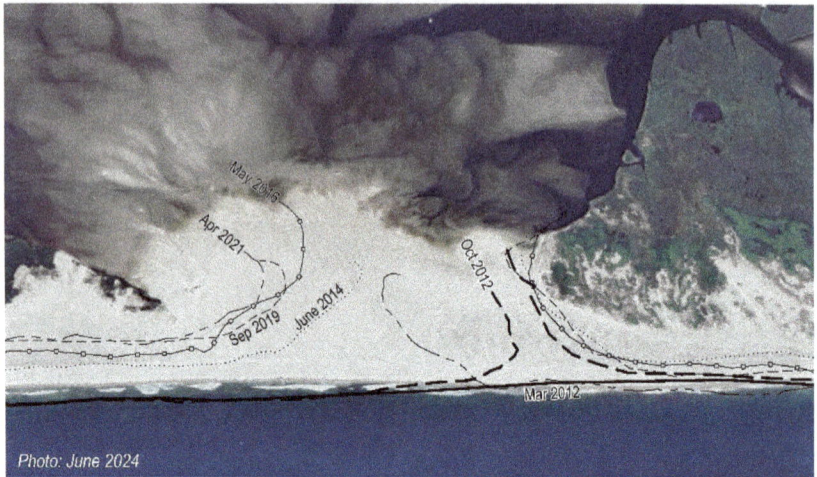

Fig. 26. Representative shorelines (waterline) at the Wilderness Breach superimposed on a June 2024 image. Note the recent shoreline nearly matches the pre-breach shoreline (May 2012) indicating little impact on the strandline after ten years of sand losses to the flood tidal delta. (Image source: Google Earth™).

and displaced most of the volume to the bay shoals (flood tidal delta). The associated ebb–tidal delta remained small but present throughout the period. The inlet was in unstable scour mode for about six years, widening and migrating west before it began to shoal from both sides and close (unstable shoaling mode) (Fig. 16). Despite the high loss of sand to the bay shoals, after the inlet closed, the ocean shoreline returned to its pre-breach condition and location.

The south shore of Long Island typically exhibits an offshore bar and associated rhythmic topography (sand waves) at 0.75–1 km spacing along the beach (Thevenot and Kraus, 1995). This produces variations in berm width of the order 40–50 m alongshore. As these sand waves propagate alongshore, they add or subtract upwards of 100 m^3/m in cross-barrier volume. In wave-dominated, sand-rich settings such as Long Island and Hatteras Island, alongshore mobility of these large sediment packages, in the author's opinion, is a key factor in breach vulnerability (higher in the berm troughs) and closure tendency (greater at the sand wave crests). The relatively high wave energy in the Outer Banks and Fire Island combined with medium to coarse sand sizes on the beach and steeper foreshore slopes lead to higher wave runup and higher berm elevations. These factors tend to accelerate breach closure, particularly compared with lower energy settings where the sediments are finer and foreshore slopes and berm elevations lower (e.g., Edingsville Beach, SC, and the Chandeleur Islands, LA).

5. Summary and Conclusions

The examples of breach inlets discussed herein occurred over a representative range of US East Coast barrier island settings with differing vulnerabilities to breaching. It is by no means a comprehensive review of breach processes and the hydrodynamics and sediment dynamics of inlet formation. These topics have been addressed by many researchers, and there are ongoing efforts to develop better predictive models (e.g., Nienhuis *et al.*, 2021; Van Ormondt *et al.*, 2020). However, the point of reviewing some specific events is to place them in context and illustrate fundamental differences in how certain barrier islands are more susceptible to breaching. The settings reviewed range from wave-dominated, microtidal barrier islands (e.g., Fire Island, NY, and the Outer Banks, NC) to the mixed energy coast of South Carolina and the rapidly subsiding coast of Louisiana. With strong evidence for accelerated SLR, the question raised at the beginning of this chapter was as follows: Will places such as the Outer Banks experience wholesale disintegration of barrier islands like many sites along Louisiana's rapidly sinking coast? This is a key question for coastal zone managers and property owners — one that needs rational answers to support sound policies, at least over decadal to century time frames.

The conventional wisdom is that breach inlets form under ocean surge conditions that overtop a low, narrow barrier beach and cut away a channel before the storm subsides. Further cutting, critical for many complete breaches, occurs as the lagoon tide ebbs through the new cut. Basco and Shin (1999) were among the first researchers to simulate this process using a coupled 1-D model (see their paper for details). They demonstrated the importance of the height difference between the ocean surge and bay (lagoon) surge, a necessary prerequisite for the hydraulic force needed to cut a breach channel to below MLLW in the short time a storm abates. They also determined that the initial dune erosion phase is less important than the channel-cutting phase. From a volume transport standpoint, this makes sense because dunes typically represent a small fraction of the subaerial volume across a barrier island. The larger volume between normal washover elevations and mean lower low water at a site has to be removed to produce a new channel.

The phase lag between the ocean and bay tide is a critical parameter because it controls the duration of channel cutting. Small or negligible phase lags, such as the situation under full barrier overtopping or sites with nearby multiple inlets, are less likely to produce a breach. Return flows will seek existing channels as the lagoon surge subsides.

Mixed energy settings such as South Carolina, where tidal and wave energy at inlets are in closer balance, more easily maintain hydraulically stable channels. Closely spaced inlets, particularly ones that share the lagoon tidal prism, provide efficient flow pathways, lessening the need for a breach channel to create a shortcut for lagoon tides. Only two breaches have occurred along the developed shoreline of South Carolina in the past 80 years (Fig. 1) despite the fact that a number of developments are situated on low, narrow barrier spits. With a 2 m tide range in the lagoons as well as the ocean, even small channels persist on that coast.

The accommodation space for breach deposits is a critical parameter. Many SC and southern NC lagoons are marsh-filled (Fig. 5). These areas may be fronted by rapidly eroding transgressive barrier beaches without new inlets forming. Masonboro Island, NC, is one such barrier that has not been breached in nearly a century despite its sand deficit. In contrast, washover barriers such as Cape Romain, SC, which is backed by an open lagoon, are prone to breaching due to a lack of sand inputs and exceedingly low subaerial dimensions (Fig. 6). The high tide range and hydraulic connection with nearby inlets reduces the ocean-bay phase lag and maintains breach channels. This is similar to conditions in rapidly subsiding — but much lower tide range — Louisiana (Fig. 3). Cape Romain is likely to be the only site on the SC coast that will experience barrier island disintegration (Kana et al., 2013).

The wave-dominated coast of the Outer Banks offers contrasting lessons in breach dynamics. This relatively sand-rich environment has only four inlets in 275 km and most of them are maintained by dredging. Mallinson et al. (2008) identified 40 breach points that have occurred in the past several centuries, nearly all having closed naturally or artificially. The wide spacing of inlets and the large size of Pamlico Sound limit the tide range to ~0.15 m. Even near the existing inlets, tides are dampened quickly. This low lagoon tide range reduces the hydraulic gradient between ocean and sound, the force needed to turn breach channels into permanent inlets.

The classic theory of Escoffier (1940) and the empirical equations of Keulegan (1967) and O'Brien and Dean (1972) help explain why most breaches in wave-dominated settings such as the Outer Banks and south shore of Long Island are likely to remain on the "unstable" side of inlet stability curves (Fig. 16). As Czerniak (1977) demonstrated for Moriches Inlet, NY, breaches may scour and expand for an initial period but are likely to shoal and close through inefficiencies in flows in low tide-range settings. At Moriches Inlet, the natural tendency to shoal and close was only overcome by periodic dredging.

Two recent hurricanes in the Outer Banks illustrate the importance of cross-sound wind tides or seiches (McNinch *et al.*, 2012). Hurricane Irene (2011) tracked up Pamlico Sound and activated some washovers or breached dunes in places. As the storm tracked north, initial winds set up water levels on the ocean side and set down levels on the sound side of the barrier islands. After the storm passed the area, winds reversed and piled up sound water against the back barrier. These "wind tides" were higher than the ocean surge particularly where there was a long fetch (order of 40 km). Breaches occurred from the sound side to release this water across low areas of the barrier. Hurricane Dorian (2019) caused numerous breaches from the sound side where the lagoon is wide but no breaches of a washover barrier — Core Banks — where the sound is much narrower (see Section 2.2). This breach process is likely to correlate with lagoon width but not be as important where lagoons are marsh-filled. The latter won't have as much water available to set up at the downwind end of a fetch. This cross-barrier wind tide effect may not be as important for events where a storm tracks directly into the coast. However, such direct tracks of storms will generally produce higher peak surges, as was the case at Westhampton Beach during the 1938 hurricane (Fig. 2). All this is to say that the breach problem is multifaceted and will be controlled by numerous site-specific factors, challenging any modeling efforts.

From a coastal zone management perspective, the main questions are as follows: Where will a breach occur, and how can events be mitigated?

High dunes and wide beaches are fundamental to shore protection along barrier islands. Wide beaches feed dunes and help maintain elevation and a reservoir of sand to accommodate seasonal changes in the littoral profile. Even when dunes are washed out by a catastrophic storm surge, they serve as speed bumps, add volume to washovers, and increase the time needed to scour a channel. Until fully predictive models are available, coastal managers can use barrier cross-sections as a proxy for breach vulnerability. This is necessarily site-specific, but recent experience demonstrates that barriers in wave-dominated settings such as Fire Island or Hatteras Island are more likely to breach if the subaerial volumes fall below 400 m³/m. This threshold value applied to breaches at Westhampton Beach (1962, 1992) and Old Inlet on Fire Island (2012). Loss of barrier volume can only be mitigated via additions of sand. It is not uncommon today for beach nourishment to involve fill densities of 100–200 m³/m. Such volumes go a long way toward improving resiliency and lessening the possibility of a breach.

The question posed at the beginning of this chapter was whether there will be wholesale disintegration of barrier islands such as the Outer Banks under present projections of SLR and increased storm intensity over the next century, or so. The experience with breaches in wave-dominated, sand-rich settings such as Hatteras Island or Fire Island suggests this is exceedingly unlikely. If disintegration is to occur by 2100, it would likely impact Core Banks, NC, first. Core Banks, despite being a washover barrier backed by an open lagoon, has not experienced wholesale breaching. Even an event that produced dozens of channels saw the breaches fully healed in less than two years (Fig. 9). There is a good reason why barrier islands in high wave, sand-rich settings are long and continuous with widely spaced inlets (Hayes, 1979). Waves build up bars wherever there is excess sand in shallow water above the local depth of closure (Kana, 2002). As long as the incident wave base is deeper than the lagoon protected by the barrier, sediments will accumulate as bars and simply be displaced upward and landward following the Bruun rule (Bruun, 1962), as modified by Rosati *et al.* (2013) to account for washover losses. Breaches will close in such settings by longshore transport from adjacent beaches just as the Wilderness Breach at Old Inlet closed (Fig. 26). Even where there are numerous breaches after a catastrophic storm (Fig. 2), intact areas will serve as feeders such that a new strand develops further landward.

The basic fallacy of long segments of the Outer Banks becoming broad inlet openings is the simple fact that the sandy sediments of the island remnants will erode more rapidly and close the gaps until a new strand line develops. Hatteras Island is likely to remain a wave-dominated, microtidal barrier island with few tidal inlets even under most likely SLR scenarios, albeit in a more landward position if no action is taken to maintain the present shoreline. But the breach experience presented herein suggests there is little chance multiple broad inlets will persist or significantly raise tides in Pamlico Sound.

One final variable that needs to be considered in much more detail by modelers and managers is sediment grain size. Profiles along the disintegrating Louisiana coast tend to be broad and shallow, with little elevation in the subaerial barriers. The dominant sediments are very fine sand ($D_{50} < 0.15$ mm) and mud leading to low relief across the foreshore. By comparison, Outer Banks and Long Island beach sands are in the medium to coarse size range ($D_{50}\sim0.5$ mm). The typical profiles of these micro tidal settings are quite different, as well (Fig. 27). Coarse, easily drained sand in the Outer Banks combined with high wave energy builds higher berms; high winds build higher dunes, and the depth in the adjacent sound is well above the local depth of closure (DOC). Sandy material in the sound is plentiful and available to be reshaped into a barrier profile if there is a major shoreline retreat in the future. By comparison, Louisiana has much lower incident waves with less capacity to build up barriers. With much less quality sand in the back barrier areas and rapid subsidence of the land, there is less chance of the barrier reconstituting itself in a

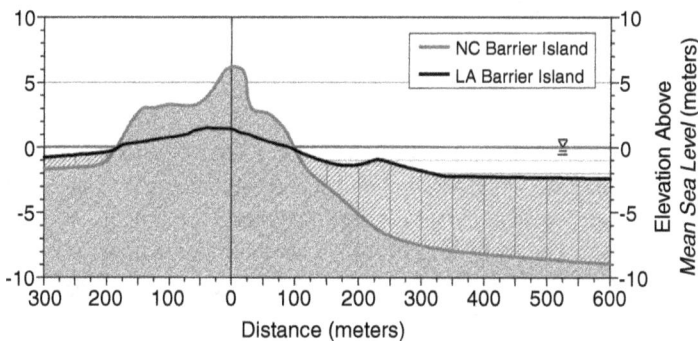

Fig. 27. Idealized barrier island cross-sections for the Outer Banks of North Carolina, where medium to coarse sand dominates, and Louisiana, where very fine sand and mud dominate. High wave energy, coarser sediments, and deeper DOC in NC account for the steeper shoreface. DOC is much shallower in Louisiana, but the volume required for a stable beach is greater (Dean, 2002). In short, less volume is needed to create a NC barrier profile for a given segment of the coast. (After Kana, 2011).

landward position. The low subaerial barrier volumes of Louisiana will continue to be vulnerable to overtopping and breaching much more frequently than the Outer Banks.

6. Acknowledgments

The examples and data presented in this chapter are based on the author's research and experience gained over the past 45 years in measuring and monitoring barrier island erosion and designing over 50 beach restoration projects in South Carolina, North Carolina, and Long Island. However, the genesis of most of the ideas goes back to many colleagues and mentors (including but not limited to those cited herein) who have worked on inlet formation problems. The goal is to highlight some important findings of others and place the problem of breach formation in context to assist coastal zone managers. Any conclusions or opinions regarding the management or interpretation of the problem are solely the author's. I thank the many municipalities, states, and funding agencies that have supported my company's applied research and erosion solutions since the 1980s.

Finally, I especially thank Prof. Young C. Kim and his reviewers for the invitation to include a chapter in this volume. Trey Hair and Carrie Marks prepared the graphics and manuscript.

References

Aubrey, D. G. and Weishar, L. (Eds.) (1988). *Lecture Notes on Coastal and Estuarine Studies* — Hydrodynamics and Sediment Dynamics of Tidal Inlets, Springer-Verlag, New York, NY, p. 454.

Aubrey, D. G. and Giese, G. S. (1993). *Formation and Evolution of Multiple Tidal Inlets*, Coastal and Estuarine Studies 44, American Geophysical Union, Washington D.C., p. 237.

Bagnold, R. A. (1941). *The Physics of Blown Sand and Desert Dunes*, Chapman and Hall, London, UK, p. 265.

Basco, D. R. and Shin, C. S. (1999). A one-dimensional numerical model for storm-breaching of barrier islands. *Journal of Coastal Research*, 15(1), 241–260.

Bocamazo, L. M., Grosskopf, W. G., and Buonuiato, F. S. (2011). Beach nourishment, shoreline change, and dune growth at Westhampton Beach, New York, 1996–2009. *Journal of Coastal Research*, Special Issue No. 59, 181–191.

Bruun, P. (1962). Sea-level rise as a cause of shore erosion. *Journal of Waterways and Harbor Div*, ASCE, New York, NY, 88(WW1), 117–132.

Bruun, P. and Gerritsen, F. (1960). *Stability of Coastal Inlets*, North Holland Publishing Company, Amsterdam, p. 140.

Cashin Assoc. (1993). *The Environmental Impacts of Barrier Island Breaching with Particular Focus on the South Shore of Long Island*, New York, Prepared for NY State Dept. of State, Albany, p. 44.

Conley, D. C. (1994). Impact of Little Pike's Inlet on tides and salinity in Moriches Bay: Final Report Marine Sciences Research Center, State University of New York, Stony Brook, NY, p. 8.

Conley, D. C. (2000). Numerical modeling of Fire Island storm breach impacts upon circulation and water quality of Great South Bay. *Continental Shelf Research*, 19, 1733–1754.

Czerniak, M. T. (1976). Engineering concepts and environmental assessment for the stabilization and sand bypassing of Moriches Inlet, New York, TetraTech Rept. to U.S. Army Corps of Eng, New York, NY, p. 102 + app.

Czerniak, M. T. (1977). Inlet interaction and stability theory verification. *Coastal Sediments*, 77 (ASCE), 754–773.

Davis, R. A. Jr. (1994). *Geology of Holocene Barrier Islands*, Springer-Berl, Berlin, p. 475.

Dean, R. G. (2002). *Beach Nourishment: Theory and Practice*, World Scientific, NJ, p. 399.

Dolan, R. and Lins, H. (1986). The Outer Banks of North Carolina, Professional Paper 1177-B, Prepared in cooperation with the National Park Service, U.S. Geological Survey, Reston, VA, p. 49.

Dolan, R., Lins, H. F., and Smith, J. J., (2016), The Outer Banks of North Carolina: U.S. Geological Survey Professional Paper 1827, p. 153, https://doi.org/10.3133/pp1827.

Donnelly, C., Kraus, N., and Larson, M. (2006). State of knowledge on measurement and modeling of coastal overwash. *Journal of Coastal Research*, 22(4), 965–991.

Doughty, S. D., Cleary, W. J., and McGinnis, B. A. (2006). The recent evolution of storm-influenced retrograding barriers in South Eastern North Carolina, USA. *Journal of Coastal Research*, SI 39, 122–126.

Escoffier, F. F. (1940). The stability of tidal inlets. *Shore & Beach*, 8(4), 114–115.

Escoffier, F. F. (1972). Hydraulics and stability of tidal inlets, U.S. Army Corps of Eng., GITI Report 13, p. 72.

Everts, C. H., Battley, J. P., and Gibson, P. N. (1983). Shoreline movements: Report 1: Cape Henry, Virginia, to Cape Hatteras, North Carolina, 1849–1980, Technical Report CERC-83-1, Coastal Engineering Research Center, U.S. Army Engineer Waterways Exp. Station, Vicksburg, MS, p. 111.

FitzGerald, D. M. (1993). Origin and stability of tidal inlets in Massachusetts. In Aubrey D. G. and Giese, G. S. (Eds.). *Formation and Evolution of Multiple Tidal Inlets*, American Geophysical Union, Washington, D.C., pp. 1–61.

FitzGerald, D. M., Buynevich, I. V., and Argow, B. (2006). Model of tidal inlet and barrier island dynamics in a regime of accelerated sea-level rise. *Journal of Coastal Research*, SI 39, 789–795.

FitzGerald, D., Howes, N., Kulp, M., Hughes, Z., Georgiou, I., and Penland, S. (2007). Impacts of rising sea level to backbarrier wetlands, tidal inlets, and barriers: Barataria Coast, Louisiana, Coastal Sediments 07, Conference Proceedings, CD-ROM13.

Flagg, C. N. and Flood, R. (2013). The Impact on Great South Bay of the Breach at Old Inlet. School of Marine and Atmospheric Sciences, Stony Brook University, p. 9.

Gaudiano, D. J., and Kana, T. W. (2001). Shoal bypassing in South Carolina tidal inlets: Geomorphic variables and empirical predictions for nine mesotidal inlets. *Journal of Coastal Research*, 17, 280–291.

GCETF. (1993). Volume 2, Long-term strategy, Draft Final Report, Governor's Coastal Erosion Task Force, Department of Environmental Conservation and Deptartment of State, Albany, NY, p. 186.

Hapke, C. J., Nelson, T. R., Henderson, R. Z., Brenner, O. T., and Miselis, J. L. (2017). Morphologic evolution of the wilderness area breach at Fire Island, New York 2012-15. Open-file Report 2017–1116. U.S. Geological Survey. p. 17 https://doi.org/10.3133/ofr20171116 (accessed January 2025).

Hayes, M. O. (1967). Hurricanes as geological agents: Case studies of hurricanes Carla, 1961, and Cindy, 1963, Report of Investigations — No 61, Bureau of Economic Geology, University of Texas, Austin, p. 56.

Hayes, M. O. (1975). Morphology of sand accumulation in estuaries. In Cronin, L. E. (Ed.), *Estuarine Research*, Academic Press, New York, Vol. 2, pp. 3–22.

Hayes, M. O. (1979). Barrier island morphology as a function of tidal and wave regime. In Leatherman, S. (Ed.), *Barrier Islands*, Academic Press, New York, pp. 1–26.

Hayes, M. O. (1980). General morphology and sediment patterns in tidal inlets. *Sedimentary Geology*, 26, 139–156.

Hayes, M. O. (1991). Geomorphology and sedimentation patterns of tidal inlets: A review. In Kraus, N. C., Gingerich, K. J., and Kreibel, D. L., (Eds.), *Coastal Sediments '91*. American Society of Civil Engineers, pp. 1343–1355.

Hayes, M. O. (1994). Georgia bight, In Chapter 7, Davis, R. A. Jr. (Ed.), *Geology of Holocene Barrier Island Systems*, Springer-Verlag, Berlin, pp. 233–304.

Hinrichs, C., Flagg, C. N., and Wilson, R. E. (2018). Great South Bay after Sandy: changes in circulation and flushing due to new inlet estuaries and coasts, Coastal and Estuaries Research Federation. Springer, p. 19.

Inman, D. and Dolan, R. (1989). The Outer Banks of North Carolina: Budget of sediment and inlet dynamics along a migrating barrier system. *Journal Coastal Research*, 5(2), 193–237.

IPCC. (2023). Climate Change 2023: Synthesis Report. Contribution of Working Groups I, II and III to the Sixth Assessment Report of the Intergovernmental Panel on Climate Change [Core Writing Team, H. Lee and J. Romero (Eds.)]. IPCC, Geneva, Switzerland, p. 184.

Irish, J. L. and Canizares, R. (2009). Storm-wave flow through tidal inlets and its influence on bay flooding. *Journal of Waterway, Port, Coastal and Ocean Engineering*, ASCE, 135(2), 52–60.

Jarrett, J. T. (1976). Tidal prism-inlet area relationships, GITI Report 33, U.S. Army Corps of Engineers Research Center, Waterways Exp. Station, Vicksburg, MS, p. 55.

Kaczkowski, H. L. and Kana, T. W. (2012). Final design of the Nags Head beach nourishment project using longshore and cross-shore numerical models, *Proceedings 33rd International Conference on Coastal Engineering* ICCE, p. 24.

Kaczkowski, H. L., Kana, T. W., Traynum, S. B., and Visser R. (2018). Beach-fill equilibration and dune growth at two large-scale nourishment sites. *Ocean Dynamics*, 68, 1191–1206.

Kana, T. W. (1995). A mesoscale sediment budget for Long Island, New York. *Marine Geology*, 126, 87–110.

Kana, T. W. (2002). Barrier island formation via channel avulsion and shoal bypassing. In Smith, J. M. (Ed.) *Proceedings 28th Internce Conf Coastal Engineering 2002*, World Scientific Publishing Co, pp. 3438–3448.

Kana, T. W. (2011). *Coastal Erosion and Solutions – A Primer*, Second Edition, Coastal Science & Engineering, Columbia, SC, p. 38.

Kana, T. W. and Mohan, R. (1994). Assessment of the vulnerability of the Great South Bay shoreline to tidal flooding, Final Report, New York Coastal Partnership, Babylon, NY, Coastal Science & Engineering, Columbia, SC, p. 153. + appendix.

Kana, T. W. and Mohan, R. K. (1996). Profile volumes as a Measure of Erosion Vulnerability. *Proceedings 25th International Conference on Coastal Engineering*, 3, 2732–2745.

Kana, T. W., Rosati, J. D., and Traynum, S. B. (2011). Lack of evidence for onshore sediment transport from deep water at decadal time scales: Fire Island, New York. *Journal of Coastal Research*, 59, 61–75.

Kana, T. W., Traynum, S. B., Gaudiano, D., Kaczkowski, H. L., and Hair, T. (2013). The physical condition of South Carolina beaches 1980–2010, *Journal of Coastal Research*, Special Issue 69, 61–82.

Keulegan, G. H. (1951). Wind tides in small closed channels. *Journal of Research of the National Bureau of Standards*, 47, 358–381.

Keulegan, G. H. (1967). Tidal flow in entrances: Water level fluctuations of basins in communication with seas, U.S. Army Corps of Engineers, Tech. Bull. 14, Committee on Tidal Hydraulics, Washington, D.C.

Kraus, N. C. and Wamsley, T. V. (2003). Coastal barrier breaching, Part 1: Overview of breaching processes: ERDC/CHL CHETN-IV-56, U.S. Army Corps of Engineers Engineer Research and Development Center, Coastal and Hydraulics Laboratory, Vicksburg, MS. p. 14.

Leatherman, S. P. (1979). Barrier dune systems: A reassessment. *Journal of Sedimentary Geology*, 24, 1–16.

Leatherman, S. P. and Joneja, D. C. (1980). Geomorphic analysis of south shore barriers, Long Island, New York: Phase I Final Report, NPS, Cooperative Research Unit, The Environmental Inst, University of Massachusetts at Amherst, p. 168.

Leffler, M., Baron, C., Scarborough, B., Hathaway, K., Hodges, P., and Townsend, C. (1996). Annual data summary for 1994 CERC Field Research Facility (2 volumes), USACE-WES, Coastal Engineering Research Center, Vicksburg, MS, Tech Rept CERC-96-6.

LIRPB. (1994). South shore mainland hazard management program, Draft Report, Long Island Regional Planning Board, Hauppauge, NY.

List, J. H., Hansen, M. E., Sallenger, A. H., and Jaffe, B. E. (1997). The impact of an extreme event on the sediment budget: Hurricane Andrew in the Louisiana Barrier Islands, *Proceedings of International Coastal Engineering Conference*. American Society of Civil Engineers, pp. 2756–2769.

Liv, J. T., Stauble, D. K., Giese, G. S., and Aubrey, D. G. (1993). Morphodynamic evolution of a newly formed tidal inlet. In Aubrey, D. G. and Giese, G. S. (Eds.), *Formation and Evolution of Multiple Tidal Inlets*. American Geophysical Union, Washington, D.C., pp. 62–94.

Lorenzo-Trueba, J. and Ashton, D. (2014). Rollover, drowning, and discontinuous retreat; Distinct modes of barrier response to sea-level rise arising from a simple morphodynamic model. *Journal of Geophysical Research: Earth Surface*, 119(4), 779–801.

Mallinson, D., Culver, S., Riggs, S., Walsh, J. P., Ames, D., and Smith, C. (2008). *Past, Present and Future Inlets of the Outer Banks Barrier Islands, North Carolina*, Department of Geophysical Sciences White Paper, East Carolina University, Greenville, NC, p. 22.

McNinch, J. E., Brodie, K. L., Wadman, H. M., Hathaway, K. K., Slocum, R. K., Mulligan, R. P., Hanson, J. L., and Birkemeier, W. A. (2012). Observations of wave runup, shoreline hotspot erosion, and sound-side seiching during Hurricane Irene at the Field Research Facility. *Shore & Beach*, 80(2), 19–37.

McSweeney, S. L., Kennedy, D. M., Rutherford, I. D., and Stout, J. C. (2017). Intermittently closed/open lakes and lagoons: Their global distribution and boundary conditions. *Journal of Geomorphology*, 292, 142–152.

Moses, R. (1939). New York metropolitan beach development. *Shore & Beach*, 7(4), 1129–1132.

Moslow, T. F. (1980). Stratigraphy of mesotidal barrier islands, Unpublished PhD Dissertation, University South Carolina, Columbia, p. 187.

Moslow, T. F. and Heron, S. D. Jr. (1994). The Outer Banks of North Carolina. In Davis, R. A. Jr. (Ed.) *Geology of Holocene Barrier Island Systems*, Springer, NY, pp. 47–74.

NCDCM. (2011). *North Carolina Long-Term Average Annual Rates of Shoreline Change: Methods Report.* North Carolina Division of Coastal Management, Department of Environment and Natural Resources, Raleigh, NC.

Nersesian, G. K., Kraus, N. C., and Carson, F. C. (1993). Functioning of groins at Westhampton Beach, Long Island, NY, *Proceedings of the 23rd Coastal Engineering Conference*, ASCE, New York, NY, pp. 3357–3370.

Nienhuis, J. H., Heijkers, L. G., and Ruessink, G. (2021). Barrier breaching versus overwash deposition: Predicting the morphologic impact of storme on coastal barriers. *Journal of Geophysical Research: Earth Surface*, 126, e2021JF006066.

NPS. (2015). Beach Restoration to Protect NC Highway 12 at Buxton, Dare County, North Carolina, Environmental Assessment, National Park Service, Cape Hatteras National Seashore, North Carolina, p. 214 + appendices.

NPS. (2017). Fire Island wilderness breach management plan – Final Environmental Impact Statement. National Park Service, Department of the Interior, Patchogue, NY, p. 188.

NPS. (2020). Sediment management framework–Cape Hatteras National Seashore, Draft Environmental Impact Statement, U.S. Department of the Interior, Manteo, NC, p. 149.

Nummedal, D., and Humphries, S. M. (1978). Hydraulics and dynamics of North Inlet, South Carolina, 1975–1976, GITI Rept. No. 16, Coastal Engineering Research Center, U.S. Army Corps of Engineers, Ft. Belvoir, VA, p. 214.

Nummedal, D., Penland, S., Gerdes, R., Schramm, W., Kahn, J., and Roberts, H. (1980). Geologic response to hurricane impact on low-profile Gulf Coast barriers, *Transactions, Gulf Coast Association of Geological Societies*, 30, 183–196.

O'Brien, M. P. (1931). Estuary tidal prisms related to entrance areas. *Civil Engineering*, 1(8), 738–739.

O'Brien, M. P. (1969). Equilibrium flow areas of inlets on sandy coasts, *Journal Waterways and Harbors Div.*, ASCE, New York, NY, 95, 43–52.

O'Brien, M. P. and Dean, R. G. (1972). Hydraulics and sedimentary stability of coastal inlets. *Proceedings of the 13th International Coastal Engineering Conference*, ASCE, New York, NY, Vol. II, pp. 761–780.

Overton, M. F. and Fisher, S. J. (2005). Bonner Bridge replacement: Parallel bridge corridor with NC 12 maintenance: Shoreline change and stabilization analysis. Prepared for URS Corporation–North Carolina and NCDOT, Task Orders 18 and 20, TPI No B-2500, FDH Engineering, Raleigh, NC, p. 39.

Pierce, J. W. (1970). Tidal inlets and washover fans. *Journal of Geology*, 78(2), 230–234.

Pilkey, O. H., Neil, W. J., Riggs, S. R., Webb, C. A., Bush, D. M., Pilkey, D. F., Bullock, J., and Cowan, B. A. (1998). *The North Carolina Shore and its Barrier Islands: Restless Ribbons of Sand*, Duke University Press, Durham, NC. p. 344.

Rather, J. (2001). Corps drops sand-replenishment plan for Fire Island. *New York Times*, 15 April 2001, Sect LI, p. 14.

Riggs, S. R. and Ames, D. V. (2003). Drowning the North Carolina coast: Sea-level rise and estuarine dynamics, Tech Rept, NCDENR, and NC Sea Grant, Raleigh, p. 154.

Riggs, S. R., Culver, S. J., Ames, D. V., Mallinson, D. J., Corbett, D. R., and Walsh, J. P. (2008). *North Carolina's Coasts in Crisis: A Vision for the Future*, Department of Geological Science White Paper, East Carolina University, Greenville, NC, p. 26.

Rosati, J. D., Gravens, M. B., and Smith, W. G. (1999). Regional sediment budget for Fire Island to Montauk Point, New York, USA. *Proceedings of Coastal Sediments '99* (New York, New York: ASCE), 802–817.

Rosati, J. D., Dean, R. G., and Walton, T. L. (2013). The modified Bruun Rule extended for landward transport. *Marine Geology*, 340, 71–81.

Ruby, C. H. (1981). Clastic facies and stratigraphy of a rapidly retreating cuspate foreland, Cape Romain, South Carolina, PhD Thesis, Department of Geology, University of South Carolina, Columbia, p. 207.

Schwab, W. C., Baldwin, W. E., Hapke, C. J., Lentz, E. E., Gayes, P. T., Denny, J. F., List, J. H., and Warner, J. C. (2013). Geologic evidence for onshore sediment transport from the inner continental shelf: Fire Island, New York. *Journal of Coastal Research*, 29, 526–544.

Sexton, W. J., and Hayes, M. O. (1996). Holocene deposits of reservoir quality sand along the central South Carolina coast. *Bulletin American Association Petroleum Geologists*, 80(6), 831–855.

Shawler, J. L., Cierletta, D. J.,Connel, J. E., Boggs, B. Q., Lorenzo-Trueba, J., and Hein, C. J. (2021). Relative influence of antecedent topography and sea-level rise on barrier island migration. *Sedimentology*, 68(2), 639–669.

Sherwood, C. R., Ritchie, A. C. Over, J. R. Kranenberg, C. J. Warrick, J., A. Brown, J. A. *et al.* (2023). Sound-side inundation and seaward erosion of a barrier island during hurricane landfall. *Journal of Geophysical Research: Earth Surface*, 128(1), 32.

Sorensen, R. M. and Schmeltz, E. J. (1982). Closure of the breach at Moriches Inlet. *Shore & Beach*, 50(4), 22–40.

Terchunian, A. V. and Merkert, C. L. (1995). Little Pike's Inlet, Westhampton, New York. *Journal of Coastal Research*, 11(3), 687–703.

Thevenot, M. M. and Kraus, N. C. (1995). Longshore sand waves at Southampton Beach, New York: Observation and numerical simulations of their movement. *Marine Geology*, 126(1–4), 249–269.

USACE. (1957). Fire Island to Jones Inlet, Long Island, NY. Cooperative Beach Erosion Study: Letter from the Secretary of the Army to Congress: House Document No. 411, U.S. Government Printing Office, Washington, D.C., p. 63.

USACE. (1958). Atlantic Coast of Long Island, New York (Fire Island Inlet to Montauk Point), Cooperative Beach Erosion Control and Interim Hurricane Study, Appendices, U.S. Army Corps of Engineers, New York.

USACE. (1963). Report on Operation Five High disaster recovery operations from 6–8 March 1962 storm, U.S. Army Corps of Engineers, North Atlantic Division, Civil Works Branch.

USACE. (1967). Atlantic Coast of Long Island, New York, beach erosion control report — Fire Island to Montauk Point, U.S. Army Corps of Engineers, House Document No. 191, Supt. Documents, Washington, D.C., p. 119.

USACE. (1996). Fire Island to Montauk Point, Long Island, New York: Breach contingency plan (executive summary and environmental assessment), U.S. Army Corps of Engineers, New York District, NY, p. 582. + appendices.

USACE. (1980). Fire Island Inlet to Montauk Point, Long Island, New York, beach erosion control and hurricane protection project, Supplement No. 2 to General Design Memorandum No. 1, Moriches to Shinnecock Reach, U.S. Army Engineer District, New York, p. 23 + 25 figure plates + 9 appendices.

USACE. (2020). Fire Island Inlet to Montauk Point, Long Island, New York, Final General Reevaluation Report, U.S. Army Engineer District, New York, p. 199.

Van Ormondt, M., Hapke, C., Roelvink, D., and Nelson, T. (2015). The effects of geomorphic changes during Hurricane Sandy on water levels in Great South Bay. *Proceedings of Coastal Sediments 2015*, p. 14. DOI: 10.1142/978981468977_0221.

Van Ormondt, M., Nelson, T. R., Hapke, C. J. and Roelvink, D. (2020). Morphodynamic modeling of the wilderness breach, Fire Island, New York. Part I Model set-up and validation. *Journal of Coastal Engineering*, 157, 103621.

Vogel, M. J., and Kana, T. W. (1985). Sedimentation patterns in a tidal inlet system, Moriches Inlet, New York, *Proceedings 19th International Coastal Engineering Conference*, ASCE, New York, NY, pp. 3017–3033.

Wainwright, D. J. and Baldock, T. E. (2015). Measurement and modeling of an artificial coastal lagoon breach. *Journal of Coastal Engineering*, 101, 1–16.

Walmsley, T. V. and Hathaway, K. K. (2004). Monitoring morphology and currents at the Hatteras Breach. *Shore & Beach*, 72(2), 9–14.

Williams, S. J., Penland, S., and Sallenger A. H. Jr. (Eds.) (1992). Louisiana barrier island erosion study-atlas of barrier shoreline changes in Louisiana from 1853 to 1989, U.S. Geological Survey Misc. Invert Service I-2150-A, p. 103.

Williams, S. J. and Foley, M. K. (2007). Recommendations for a barrier island breach management plan for Fire Island National Seashore, including the Otis Pike high dune wilderness area. Long Island, New York Technical Report, NPS/NER/NRTR-2007/075, National Park Service, Boston, MA.

Wutkowski, M. (2004). Hatteras breach closure. *Shore & Beach*, 72(2), 20–24.

Zarillo, G. A., Ward, L. G., and Hayes. M. O. (1985). *An Illustrated History of Tidal Inlet Changes in South Carolina*, South Carolina Sea Grant Consortium, Charleston, SC, p. 76.

Chapter 3

Beach Erosion and Coastal Structure Damage by Storm Surge and Waves

Nobuhisa Kobayashi*, Tingting Zhu† and Jirat Laksanalamai‡

*Center for Applied Coastal Research,
University of Delaware, Newark, DE 19716, USA.*

*nk@udel.edu

†ztting@udel.edu.

‡jiratlak@udel.edu

The majority of the world's shoreline is currently suffering from erosion. Beach erosion will become more serious if the mean sea level rise accelerates and hurricanes intensify due to global warming. Nourishment and maintenance of wide sand beaches for developed coastal communities will become unsustainable unless the present nourishment method is innovated. Concurrently, the recent increase in coastal storm damage demands the capabilities for predicting the damage progression and breaching of coastal stone structures and earthen levees during extreme storms. Effort has been made to improve our quantitative understanding of beach morphology and structural damage progression with the goal to develop simple and robust models that are suited for engineering applications. Our effort for the last 20 years has produced the cross-shore numerical model CSHORE consisting of the following components: a combined wave and current model, a probabilistic swash model, a sediment transport model for beach and dune erosion, a porous layer model

for rubble mound structure damage, a drag force model for piles inter-
acting with waves and sand dunes, a dike erosion model by irregular
wave action, wave overtopping and bay flooding, sand transport on and
inside porous structures, intertidal mudflat profile evolution, and beach
fill design. The theories and formulas used in CSHORE are explained
in order to facilitate the application of CSHORE to various engineering
problems involving coastal hydrodynamics, sediments, and structures.

1. Introduction

A sand beach with a wide berm and a high dune provides storm protection
and damage reduction, recreational and economic benefits, and biological
habitats for plants and animals. Most sandy beaches are eroding partly due
to sea level rise. Beach nourishment is widely adopted to maintain a wide
beach for a developed coastal community if a suitable beach fill is available in the vicinity of an eroding beach. Empirical methods based on field
data and an equilibrium beach profile have been developed for the design
of beach fills (USACE, 2003). The geometry of the nourished berm and
dune is designed to prevent or reduce coastal flooding by storm surges
and waves. The gradual alongshore spreading of the beach fill is generally predicted using a one-line model coupled with the CERC formula
(USACE, 2003) or the formula by Kamphuis (1991) for the longshore
sediment transport rate.

Sediment transport is caused by the combined action of waves and
currents. Our capabilities of predicting wave and current fields have
improved steadily for the last 30 years. However, the predictive capability
of sediment transport on beaches has not improved as much. The state of
the art of coastal sediment transport modeling for engineering applications
was reviewed by Kobayashi (2016). The literature on morphodynamic
modeling of coastal barriers was reviewed by Hoagland *et al.* (2023). This
review is based on the latest CSHORE documented in the report of
Kobayashi and Zhu (2022). CSHORE synthesized available data and formulas to derive simple and transparent formulas for the cross-shore and
longshore transport rates of suspended sand and bedload on beaches. The
simple formulas included basic sediment dynamics and were applicable to
small-scale and large-scale laboratory beaches as well as natural beaches.
This morphological model is very efficient computationally and easy to

apply for various coastal problems. The cross-shore model CSHORE is presently limited to the conditions of alongshore uniformity or assumed alongshore length scale for the longshore sediment transport gradient.

Coastal storm damage has been increasing mostly due to the recent growth of coastal population and assets. Coastal structures including earthen levees (dikes) and rubble mound structures have been designed conventionally for no storm surge overflow and minor wave overtopping during a design storm. Empirical formulas for wave overtopping rates are used for a preliminary design (EurOtop Manual, 2007, 2018). Physical model testing is normally conducted in a wave flume or basin for a detailed design. Various numerical models have also been developed to predict detailed hydrodynamics (Kobayashi *et al.*, 1987, 1989; Losada *et al.*, 2008; Irias Mata and van Gent, 2023; van Gent, 2001). However, our improved predictive capabilities for the hydrodynamics have not really improved our predictive capability for damage progression partly because damage to a coastal structure is cumulative and depends on damaged structure resistance (Melby and Kobayashi, 1998). As a result, a risk-based design of a coastal structure relies on empirical formulas for damage (e.g., Kobayashi *et al.*, 2003; Melby and Kobayashi, 2011). This practical difficulty is similar to that of sediment transport on beaches. The computationally efficient CSHORE calibrated with extensive datasets has also been developed for the design of inclined structures with relatively small wave reflection. Damage progression on the stone armor layer is predicted by modifying the sediment transport model (Kobayashi *et al.*, 2010a). Cohesionless sediment transport formulas should be applicable to sand, gravel, and stone.

The cross-shore model CSHORE consisting of the following components has been developed for the last two decades:

From 2003 to 2013
- Combined wave and current model,
- probabilistic swash model,
- beach and dune erosion,
- rubble mound structure damage.

From 2014 to 2024
- Effect of woody plants and piles on dune erosion and overwash,
- erosion of grassed dike and consolidated cohesive sediment,

- wave overtopping of barrier beach and bay flooding,
- sand transport on and inside porous stone structures,
- intertidal mudflat profile evolution under waves and currents,
- beach fill design under low wave energy.

The relative simplicity of CSHORE allows various extensions to examine unsolved problems in coastal engineering.

2. Combined Wave and Current Models

A combined wave and current model is adopted in CSHORE because of its computational efficiency. Sediment transport on beaches is caused by the combined action of waves and currents. The hydrodynamic input required for a sediment transport model depends on whether the sediment transport model is time-dependent (phase-resolving) or time-averaged over a number of waves. A time-dependent sediment transport model such as that by Kobayashi and Johnson (2001) is physically appealing because it predicts intense but intermittent sand suspension under irregular breaking waves (Kobayashi and Tega, 2002) in which Cox and Kobayashi (2000) measured intense, intermittent coherent motions under regular waves shoaling and spilling on a rough impermeable slope. However, the time-dependent model requires considerable computation time and is not necessarily more accurate in predicting slow morphological changes than the corresponding time-averaged cross-shore model CSHORE.

The bathymetry is assumed to be uniform alongshore. The cross-shore coordinate x is positive onshore and the longshore coordinate y is positive in the downwave direction of obliquely incident waves. The sand beach is assumed to be impermeable, and the bottom elevation z_b is positive upward with $z = 0$ at the datum. The depth-averaged cross-shore and longshore velocities are denoted by U and V, respectively. Incident waves are assumed to be unidirectional with θ as the incident angle relative to the shore normal. The height and period of the irregular waves are represented by the root-mean-square wave height H_{rms} and the representative wave period, which may be taken as the spectral peak period T_p or the spectral wave period (van Gent, 2001) specified at the seaward boundary located at $x = 0$. The location of the seaward boundary is normally taken to be outside the surf zone so that wave set-down or setup is very small at $x = 0$.

The mean water depth $\bar{h}$ is given by $\bar{h} = (\bar{\eta} + S - z_b)$ where $\bar{\eta}$ is the wave setup above the still water level (SWL) and S is the storm tide above the datum $z = 0$ which is assumed to be uniform SWL in the computation domain with a cross-shore distance of the order of 1 km and is specified as input at $x = 0$ as a function of storm tide time t. Linear wave and current theory for wave refraction (e.g., Phillips, 1977; Mei, 1989; Dalrymple, 1988) is used to predict the spatial variations of H_{rms} and θ as a function of time t. Snell's law is used to obtain the wave angle θ.

The depth-integrated time-averaged continuity equation of water requires that the cross-shore volume flux is constant and equal to the wave overtopping rate q_o at the landward end of the computation domain. The onshore volume flux induced by waves is proportional to the square of H_{rms} and is much larger than q_o except in the zone of very small H_{rms}. The mean cross-shore velocity $\overline{U}$ associated with the return flow $\overline{h}\overline{U}$ is negative and offshore. The volume flux of a roller on the front of a breaking wave is included if the roller option is selected. The cross-shore volume flux associated with the temporal variation of the still water level S may be included to add the cross-shore tidal current if the tidal range is very large and the bottom slope is very gentle (Do *et al.*, 2016).

The cross-shore and longshore momentum equations including the wave-induced radiation stresses (e.g., Phillips, 1977) are used to predict the cross-shore variations of the wave setup $\overline{\eta}$ and the longshore current $\overline{V}$ induced by breaking waves. The momentum flux of a roller propagating with the phase velocity C (Svendsen *et al.*, 2002) may be included in the momentum equations to improve the agreement for the longshore current on sand beaches (Kobayashi *et al.*, 2007a). The time-averaged bottom shear stresses in the momentum equations are written using the cross-shore velocity U, longshore velocity V, and bottom friction factor f_b. Cox *et al.* (1996) measured the velocities inside the bottom boundary layer for regular waves shoaling and spilling on a rough impermeable 1/35 (vertical/horizontal) slope. The bottom friction factor f_b of the order of 0.01 on sand beaches has been assumed, but it should be calibrated using longshore current data because of the sensitivity of longshore currents to f_b (Kobayashi *et al.*, 2007a). The equivalency of the time and probabilistic averaging is assumed to express the time-averaged bottom shear stresses in terms of the mean and standard deviation of the velocities U and V. Linear progressive wave theory is used locally to express the oscillatory depth-averaged velocity in terms of the oscillatory free surface elevation which is assumed Gaussian with the mean $\overline{\eta}$ and the standard deviation σ_η, where $H_{rms} = \sqrt{8}\sigma_\eta$ is assumed. The spectral significant wave height H_{mo} is given by $H_{mo} = 4\sigma_\eta$. The use of linear wave theory in the mean depth $\bar{h}$ for the wave angle θ leads to the simple expressions of the standard deviations of σ_u and σ_v of the velocities U and V in terms of σ_η, $\bar{h}$, and θ.

The root-mean-square wave height H_{rms} is obtained using the wave action or energy equation for the case of the wave overtopping rate $q_o > 0$ or $q_o = 0$, respectively. The energy dissipation rate D_B due to wave breaking is estimated using the simple formula by Battjes and Stive (1985), which was modified by Kobayashi *et al.* (2005) to account for the local bottom slope and to extend the computation to the lower swash zone. The breaker ratio parameter γ for the local depth-limited wave height is typically in the range of $\gamma = 0.5$–1.0 (Kobayashi *et al.*, 2007a) but should be calibrated to obtain a good agreement with the measured cross-shore variation of H_{mo} if such data are available. The typical value of γ is 0.7 (0.6 for a very gentle slope). Irregular wave breaking is very complex, and constant γ is a crude approximation. On the other hand, the energy dissipation rate D_f due to bottom friction is obtained by the probabilistic averaging of the product of the instantaneous bottom shear stress and water velocity. If the roller effect is included in the CSHORE computations, the roller energy equation is solved to obtain the cross-shore variation of the roller volume flux where the breaking wave energy dissipation rate is assumed as the generation rate of the roller energy. The roller effect creates the landward shift of the wave energy dissipation and the longshore current $\bar{V}$ (Kobayashi *et al.*, 2007a).

3. Probabilistic Swash Model

The combined wave and current model is limited to the wet zone where the mean water depth $\bar{h}$ is positive. A separate hydrodynamic model is required to predict the mean and standard deviation of the free surface elevation η and the depth-averaged velocities U and V. Time-dependent numerical models, such as the nonlinear shallow-water wave model by Kobayashi *et al.* (1987, 1989), can predict the water depth and horizontal velocity in the intermittently wet and dry (swash) zone on beaches and inclined structures. However, the time-dependent hydrodynamic computation requires considerable computation time and may not lead to an accurate prediction of beach profile evolution. Kobayashi *et al.* (2010b) developed a time-averaged probabilistic model to predict the cross-shore variations of the wet probability and the mean and standard deviation of the water depth and velocity in the swash zone. The developed model is very efficient computationally and can be calibrated using a large number of datasets. Van Gent (2002) and Schüttrumpf and Oumeraci (2005) measured and analyzed the water depth and velocity of waves overtopping of dikes. Kobayashi *et al.* (2010b) expanded their analyses for the prediction of wave overtopping and overwash as explained in the following.

For normally incident waves on impermeable beaches and inclined structures of alongshore uniformity, the time-averaged cross-shore continuity and momentum equations derived from the nonlinear shallow-water wave equations are used in the swash model in CSHORE. The corresponding wave energy equation given by Kobayashi and Wurjanto (1992) is not used because no formula is available to estimate the time-averaged energy dissipation rate associated with the interaction of wave uprush and downrush in the swash zone. The instantaneous water depth h depends on the cross-shore coordinate x and the swash hydrodynamic time t. The water depth h at given x is described probabilistically rather than in the time domain. Kobayashi *et al.* (1998) analyzed the probability distributions of the free surface elevations measured in the shoaling, surf, and swash zones. The measured probability distributions were shown to be in agreement with the exponential gamma distribution, which reduces to the Gaussian distribution and the exponential distribution when the skewness approaches zero offshore and two in the swash zone, respectively. The assumption for the Gaussian distribution has simplified CSHORE in the wet zone significantly. The assumption of the exponential distribution is made in the wet and dry zone. The probability density function $f(h)$ is expressed using the wet probability P_w for the water depth $h > 0$ and the mean water depth $\bar{h}$ for the wet duration. The dry probability of $h = 0$ is equal to $(1 - P_w)$. The mean water depth for the entire duration is equal to $P_w\bar{h}$. The free surface elevation $(\eta - \bar{\eta})$ is equal to $(h - \bar{h})$. The standard deviations of η and h are the same and the ratio of $\sigma_\eta/\bar{h}$ depends on P_w only in this probabilistic swash model (Kobayashi *et al.*, 2010b).

The cross-shore velocity U is related to the depth h in the swash zone in terms of $\alpha\sqrt{gh} + U_s$, where α is the positive constant exceeding unity for supercritical flow and U_s is the steady velocity which is allowed to vary with x. The steady velocity U_s is intended to account for offshore return flow on the seaward slope and the downward velocity increase on the landward slope. Holland *et al.* (1991) measured the bore speed and flow depth on a barrier island using video techniques and obtained $\alpha = 2$ where the celerity and fluid velocity of the bore were assumed to be approximately the same. The calibrated value by Figlus *et al.* (2012) for wave overtopping of sand dunes was $\alpha = 1.6$. The mean $\bar{U}$ and standard deviation σ_U of the cross-shore velocity U can be expressed in terms of $\bar{h}$, P_w, and U_s. Probabilistic averaging of the continuity equation yields an equation for U_s as a function of $\bar{h}$ and P_w for given wave overtopping rate q_o. The cross-shore momentum equation is averaged probabilistically to obtain a differential equation for $\bar{h}$ with respect to x. The functional form of P_w as a function of $\bar{h}$ is assumed for the integration of the differential equation, where P_w decreases with the decrease of $\bar{h}$ in the swash zone. The integrated equation is solved iteratively to compute $\bar{h}$ as a function of x. The wave overtopping rate q_o is predicted by imposing $U_s = 0$ at the crest of a dune or structure so that the cross-shore velocity U is positive and onshore at the crest. On the slope landward of the crest, the wet probability P_w is assumed to be constant and equal to the computed wet probability at the crest.

For assumed q_o, the landward marching computation of $\bar{h}$, σ_η, $\bar{U}$, σ_U, $\bar{V}$, and σ_V is initiated using the wet model from the seaward boundary $x = 0$ to the landward limit located at $x = x_r$, which corresponds to the location where the computed $\bar{h}$ or σ_η becomes negative or $\bar{h}$ becomes less than 0.1 cm for an emerged crest. For a submerged crest, the landward limit of x_r is the landward end of the computation domain with no wet and dry zone. The landward marching computation is continued using the swash (wet and dry) model from the location of the still water shoreline x_{SWL} to the landward end of the computation domain or until the mean depth $\bar{h}$ becomes less than 0.001 cm. Then, the rate q_o is computed at the crest. This landward computation starting from $q_o = 0$ is repeated until the difference between the computed and assumed values of q_o is less than 1%. This convergency is normally obtained after several iterations. The computed values of $\bar{h}$, σ_η, $\bar{U}$, and σ_U by the two different models in the overlapping zone of $x_{SWL} < x < x_r$ are averaged to smooth the transition from the wet zone to the swash zone.

Kobayashi *et al.* (2010b) compared this hydrodynamic model for the impermeable swash zone with their 107 tests of wave overtopping on an impermeable smooth levee and the 100 tests conducted by van Gent (2002) who measured the water depth and velocity on the crest and landward (inner) slopes of six different dikes. The agreement was mostly within a factor of two for the wave overtopping rates and the water depth, velocity, and discharge on the crest and landward slope exceeded by 2% of the incident 1000 waves. On the other hand, Farhadzadeh *et al.* (2012)

extended CSHORE to allow oblique waves in the swash zone for the small incident wave angle. The mean $\overline{V}$ and standard deviation σ_V of the longshore velocity V in the wet and dry zone are expressed using $\overline{h}$, P_w, and θ_1 where the wave angle θ_1 at $x = x_{SWL}$ is assumed to satisfy $(\sin\theta_1)^2 \ll 1$ and the equations for $\overline{h}$ and P_w remain the same as those for $\sin\theta_1 = 0$.

The hydrodynamic model in CSHORE has been shown to predict the cross-shore variations of the mean and standard deviation of η, U, and V within errors of about 20% (Kobayashi *et al.*, 2005; 2007a, 2007b; 2009). However, it may be necessary to calibrate the bottom friction factor f_b and the breaker ratio parameter γ. Kobayashi *et al.* (2013b) compared the wave runup formulas in CSHORE with 137 impermeable dike tests (van Gent, 2001) and 120 uniform slope tests (Mase, 1989). The measured 2% and 1% exceedance runup heights for the 257 tests were predicted within errors of about 20%. Melby *et al.* (2012) compared extensive wave runup data on natural beaches. Wave runup on dissipative beaches may not be predicted well by CSHORE when infragravity waves dominate shoreline oscillations.

4. Beach and Dune Erosion

The spatial variations of the hydrodynamic variables used in the following cohesionless sediment transport model are computed for a given beach profile, water level, and seaward wave conditions at $x = 0$. The bottom sediment is assumed to be uniform and characterized by d_{50} which is the median diameter, w_f which is the sediment fall velocity, and s which is the sediment-specific gravity. The sediment transport model consists of bed load and suspended load in the wet zone and swash zone.

The spatial variation of the degree of sediment movement is estimated using the critical Shields parameter ψ_c (Madsen and Grant, 1976), which is taken as $\psi_c = 0.05$. The sediment movement is assumed to occur when the instantaneous bottom shear stress exceeds the critical shear stress. The probability P_b of sediment movement (Kobayashi *et al.*, 2008b) can be expressed analytically for the Gaussian velocity distribution in the wet zone. The value of P_b computed from $x = 0$ located outside the surf zone increases landward and fluctuates in the surf and swash zones, depending on the presence of a bar or a terrace that increases the local fluid velocity. The spatial variation of the degree of sediment suspension is estimated using the experimental finding of Kobayashi *et al.* (2005) who showed that the turbulent velocities measured in the vicinity of the bottom were related mostly to the energy dissipation rate due to bottom friction. The probability P_s of sediment suspension has been assumed by Kobayashi *et al.* (2008b) to be the same as the probability of the estimated turbulent velocity exceeding the sediment fall velocity w_f. If $P_s > P_b$, use is made of $P_s = P_b$ assuming that sediment suspension occurs only when sediment movement occurs. Fine sands on beaches tend to be suspended once their movement is initiated. Armor stone may not be suspended even under large waves.

The suspended sediment volume V_s per unit horizontal bottom area is estimated by modifying the time-dependent sediment suspension model by Kobayashi and Johnson (2001) where V_s increases with the increase of the energy dissipation rates due to wave breaking and bottom friction. The breaking wave suspension efficiencies were calibrated using the measured V_s (Kobayashi *et al.*, 2007a). The sediment suspension probability P_s is added in the formula to ensure $V_s = 0$ if $P_s = 0$. The cross-shore and long-shore suspended sediment transport rates q_{sx} and q_{sy} by the cross-shore current $\overline{U}$ and longshore current $\overline{V}$ are expressed as $q_{sx} = a_x \overline{U} V_s$ and $q_{sy} = \overline{V} V_s$. The parameter a_x accounts for the onshore suspended sediment transport due to the positive correlation between the time-varying cross-shore velocity and suspended sediment concentration. The value of a_x increases to unity as the positive correlation decreases to zero. For the three small-scale equilibrium profile tests conducted by Kobayashi *et al.* (2005), a_x was of the order of 0.2. The cross-shore suspended sediment transport rate q_{sx} is negative (offshore) because the return (undertow) current $\overline{U}$ is negative (offshore). Madsen *et al.* (1994) measured offshore suspended sediment transport on a natural beach during a storm. On the other hand, the longshore suspended sediment transport rate q_{sy} neglects the correlation between the time-varying longshore velocity and suspended sediment concentration, which appears to be very small if the longshore current $\overline{V}$ is sufficiently large. Payo *et al.* (2009) verified the suspended load formulas using velocities and sand concentrations measured along 20 transects at the Field Research Facility at Duck, North Carolina, during a storm in 1997.

The formulas for the cross-shore and longshore bedload transport rates q_{bx} and q_{by} were devised somewhat intuitively because bedload in the surf zone has never been measured. The time-averaged rates q_{bx} and q_{by} based on a quasi-steady application of the formula of Meyer-Peter and Mueller (e.g., Ribberink, 1998) were adjusted empirically to be consistent with the onshore bedload formula proposed by Kobayashi *et al.* (2008b) for normally incident waves, which synthesized existing data (Ribberink and Al-Salem, 1994; Dohmen-Janssen and Hanes, 2002; Dohmen-Janssen *et al.*, 2002; Masselink *et al.*, 2007) and simple formulas (Bagnolds, 1966; Trowbridge and Young, 1989). The bedload transport rates q_{bx} and q_{by} are multiplied by the sediment movement probability P_b to account for the initiation of sediment movement. The condition of $(q_{bx} + q_{sx}) = 0$ for an equilibrium profile along with additional assumptions can be shown to yield the equilibrium profile popularized by Dean (1991). Kobayashi *et al.* (2018) extended the equilibrium profile model to include net cross-shore sand transport and estimate the seaward shift of the shoreline on periodically nourished beaches. To reproduce the 221-day recovery of eroded beaches after a severe storm, Kobayashi and Jung (2012) adjusted the bedload parameter $b = B(0.5 + Q)$ with $B = 0.002$ and Q as the fraction of irregular breaking waves where $b = 0.001$ outside the surf zone ($Q = 0$) was increased to $b = 0.003$ near the shoreline ($Q = 1$). This adjustment suggests the bedload uncertainty of a factor of 2.

The sediment transport model was extended to the wet and dry zone (Kobayashi *et al.*, 2010b). The Gaussian velocity distribution has been assumed in the wet zone, whereas the horizontal velocity U in the wet and dry zone is expressed as a function of the depth h along with the exponential probability distribution of h. The movement of sediment particles is assumed to occur when the instantaneous bottom shear stress exceeds the critical shear stress, as has been assumed in the wet zone. The probability P_b of sediment movement can be expressed analytically where the upper limit of P_b is the wet probability P_w because no sediment movement occurs during the dry duration. Sediment suspension is assumed to occur when the instantaneous turbulent velocity of the energy dissipation rate due to bottom friction exceeds the sediment fall velocity w_f. The probability P_s of sediment suspension is expressed analytically for computational efficiency. If $P_s > P_b$, use is made of $P_s = P_b$ because sediment suspension occurs only when sediment movement occurs.

The suspended sediment volume V_s per unit horizontal bottom area in the wet and dry zone is estimated so that the suspended sediment volume V_s is continuous at $x = x_{SWL}$ between the wet zone and the wet and dry zone. This intuitive assumption was necessary because no formula is available for sediment suspension in the swash zone. Kobayashi *et al.* (2010b) estimated the cross-shore suspended sediment transport rate q_{sx} using the formula developed for the case of no wave overtopping of a berm or dune. However, the formula was found to underpredict major wave overwash in the three small-scale tests conducted by Figlus et al. (2011) who investigated the transition from minor to major wave overwash of dunes constructed of fine sand. For these tests, suspended load was computed to be dominant. To include the effect of the wave overtopping rate q_o on the cross-shore suspended sediment transport rate q_{sx}, Figlus et al. (2011) added the effect of onshore current $U_o = q_o/\bar{h}$ due to the wave overtopping rate q_o, which is significant only in the zone of the very small depth $\bar{h}$. The degree of the U_o effect on sand overwash was larger in the data of Figlus et al. (2011, 2012) than in the field data with minor overwash used in Kobayashi et al. (2010b). The accurate prediction of wave overtopping and overwash is very difficult because of the small water depth and large velocity in the zone which is wet intermittently. The cross-shore bedload transport rate q_{bx} is estimated using the formula for q_{bx} in the wet zone where the values of q_{bx} in the two zones are matched at the cross-shore location of $x = x_{SWL}$.

Finally, the beach profile change is computed using the continuity equation of the bottom sediment

$$(1-n_p)\frac{\partial z_b}{\partial t} + \frac{\partial q_x}{\partial x} + \frac{\partial q_y}{\partial y} = 0, \tag{1}$$

where n_p is the porosity of the bottom sediment which is normally taken as $n_p = 0.4$, t is the slow morphological time for the change of the bottom elevation z_b, $q_x = (q_{sx} + q_{bx})$ is the cross-shore total sediment transport rate, and $q_y = (q_{sy} + q_{by})$ is the longshore total sediment transport rate. For the case of alongshore uniformity, the third term in Eq. (1) is zero. Equation (1) is solved using an explicit Lax–Wendroff numerical scheme (e.g., Nairn and Southgate, 1993) to obtain the bottom elevation at the next time level. This computation procedure is repeated starting from the initial bottom profile until the end of a profile evolution.

Kobayashi and Jung (2012) expanded CSHORE to allow the simultaneous computation of the multiple cross-shore lines and included the effect of the alongshore gradient of q_y in Eq. (1) on the temporal variation of z_b along each line in an approximate but computationally efficient manner. The expanded CSHORE was compared with natural beach erosion and recovery data along 16 cross-shore lines spanning 5 km alongshore for the duration of 272 days. CSHORE predicted beach and dune erosion fairly well and reproduced the accreted beach profile with a berm after the adjustment of the bedload parameter by a factor of 2.

Figlus *et al.* (2012) conducted a laboratory experiment on the onshore migration of an emerged ridge and a ponded runnel because the onshore migration of the ridge and runnel can have a significant influence on the sediment budget and beach recovery after a storm. The experiment was focused on the effect of water ponding and runnel drainage on the onshore ridge migration. The test scenario with a drained runnel showed a ridge migration speed five times larger than the other scenario in which water and sediment could only exit the runnel as offshore return flow over the ridge. CSHORE was modified to predict the ponded water level in the runnel and estimate the reduced bedload and suspended sediment transport rates in the ponded water zone. This modification enabled CSHORE to reproduce the observed sand transport asymmetry between onshore transport into the runnel and offshore transport out of the runnel where this asymmetry resulted in the deposition at the seaward end of the runnel. Measured hydrodynamics, wave overtopping, and sediment overwash rates were predicted reasonably well in light of the strong interaction between the hydrodynamics and morphological evolution.

For small incident wave angles with $(\sin \theta_1)^2 \ll 1$ in the swash zone, the cross-shore sediment transport rates q_{sx} and q_{bx} are estimated using the formulas developed for normal incidence ($\theta = 0$). The longshore suspended sediment transport rate q_{sy} is predicted using $q_{sy} = \overline{V} V_s$ in both wet and swash zones. The longshore bedload transport rate q_{by} is estimated using the formula in the wet zone with the assumption of $(\sin \theta_1)^2 \ll 1$ where the mean $\overline{V}$ and standard deviation σ_V of the longshore velocity V in the swash zone are of the order of $\sin \theta$ (Farhadzadeh *et al.*, 2012).

The effects of seawalls on sand beaches were investigated by a number of researchers. Saitoh and Kobayashi (2012) conducted an experiment to examine cross-shore sand transport on a sloping beach in front of a vertical wall located well above the SWL. An initial semi-equilibrium beach accreted slightly above SWL in which approximately 5% of incident waves reached the vertical wall. The cross-shore variations of the suspended sand and bed load transport rates predicted by CSHORE were examined to explain the net onshore sand transport, but CSHORE could not reproduce the slight accretion adequately. Johnson *et al.* (2012) evaluated CSHORE using severe beach erosion data at seven sites on the mid-Atlantic east coast and two sites on the southern California coast.

The degree of agreement for the seven mid-Atlantic sites was similar to the other comparisons discussed in this paper. The transition from no or minor overwash to major overwash of a dune is difficult to predict consistently because of the uncertainty of the overwash process. For the two California sites with gentler beach slopes and smaller storm surges, fairly good agreement was obtained by increasing the computed offshore suspended sediment transport rate in which the suspension efficiency and load parameter were increased by a factor of about 2. Kalligeris *et al.* (2020) compared CSHORE and XBEACH (Roelvink *et al.*, 2009) with six energetic wave events at two southern California beaches. Cohn *et al.* (2021) compared CSHORE and XBEACH with variable dune profile changes during weeks of elevated wave and water level conditions at a southeastern Atlantic beach. The degrees of calibration and agreement for CSHORE were similar to the previous comparisons. XBEACH required more calibrations for various erosion events and may not be applicable to accretional events. Saunders *et al.* (2024) applied CSHORE on an annual scale to a multi-barred, dissipative beach at a northwestern Pacific beach. CSHORE predicted physically realistic bar behavior, including bar growth, migration, and decay.

5. Rubble Mound Structure Damage

The preliminary design of rubble mound structures is based on empirical formulas derived from laboratory experiments on wave runup, reflection, overtopping, and transmission as well as armor layer stability. Kobayashi (2015) summarized the progress of the coastal structure design since the publication of the Coastal Engineering Manual (USACE, 2003). Hydraulic model testing is required to combine the different design elements and optimize the entire structure design. CSHORE can also be used to examine the hydraulic response and armor damage in integrated manners.

The combined wave and current model in CSHORE has been extended to allow the presence of a permeable layer in the wet zone because the permeability effect is not negligible for gravel beaches and stone structures. Figure 1 shows an example of irregular wave overtopping of a permeable slope where x is the onshore coordinate, z is the vertical coordinate, $\bar{\eta}$ is the mean free surface elevation above SWL, S is the storm tide above $z = 0$, z_b is the bottom elevation, $\bar{h}$ is the mean water depth, U is the instantaneous depth-averaged cross-shore velocity above

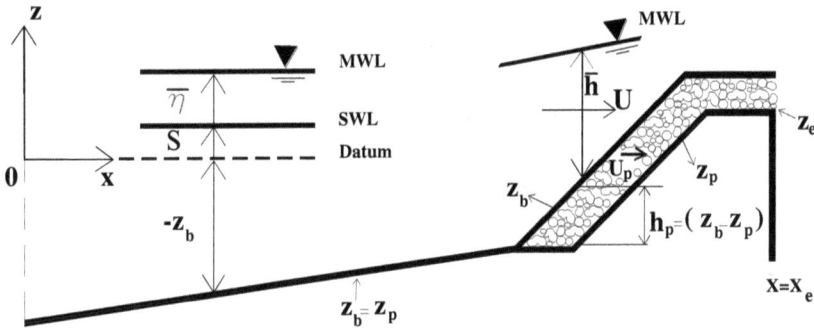

Fig. 1. Definition sketch of permeable layer model.

the bottom, z_p is the elevation of the lower boundary of the permeable layer, $h_p = (z_b - z_p)$ is the vertical thickness of the permeable layer, and U_p is the instantaneous cross-shore discharge velocity inside the permeable layer. The cross-shore profiles of $z_b(x)$ and $z_p(x)$ are specified as input where $h_p = 0$ in the zone of no permeable layer. The lower boundary located at $z = z_p$ is assumed to be impermeable for simplicity. Kobayashi *et al.* (2007b) developed a permeable layer model in the wet zone for normal incident waves. This model was extended to obliquely incident waves and was applied by Garcia and Kobayashi (2015) to predict oblique wave transmission over and through low-crested porous breakwaters (Burcharth *et al.*, 2006).

The time-dependent model for the flow over a permeable layer in shallow water developed by Kobayashi and Wurjanto (1990) and Wurjanto and Kobayashi (1993) is time-averaged and simplified to account for the permeable layer in CSHORE. The vertically integrated continuity equation in the wet zone is modified to include the time-averaged cross-shore discharge velocity $\overline{U}_p$ and the water flux $\left(h_p \overline{U}_p\right)$ inside the permeable layer with its vertical thickness h_p. The rate q_o is the combined wave overtopping rate above and through the permeable layer. The cross-shore and longshore momentum equations are assumed to remain the same, neglecting the momentum fluxes into and out of the permeable layer in the wet zone which is saturated with water. The bottom friction factor f_b including the effect of the surface roughness of the permeable layer was calibrated in the range of $f_b = 0.01$–0.05 (Kobayashi *et al.*, 2007b, 2008a). For the case of alongshore uniformity and negligible momentum fluxes into and out of the permeable layer in the wet zone, the time-averaged longshore discharge velocity $\overline{V}_p$ is assumed to be zero because of no or negligible driving force to cause the longshore discharge inside the permeable layer.

On the other hand, the wave energy or action equation is modified to include the energy dissipation rate D_p due to flow resistance in the permeable layer. The dissipation rate D_p is expressed using laminar and turbulent flow resistance coefficients (Wurjanto and Kobayashi, 1993). The formulas for the resistance coefficients by van Gent (1995) were modified for CSHORE (Kobayashi *et al.*, 2007b). The nominal stone diameter D_{n50} is defined as $D_{n50} = (M_{50}/\rho_s)^{1/3}$ with M_{50} as the median stone mass and ρ_s as the stone density. The discharge velocities U_p and V_p are assumed to be Gaussian variables and expressed in terms of the mean $\overline{U}_p$ and standard deviation σ_p of the oscillatory velocity inside the permeable layer under the incident wave angle θ. The time-averaged rate D_p is obtained by averaging the instantaneous rate probabilistically. Neglecting the inertia terms in the cross-shore momentum equation for the flow inside the permeable layer (Kobayashi and Wurjanto, 1990), the local force balance between the cross-shore hydrostatic pressure gradient and flow resistance is assumed to derive approximate equations for the mean $\overline{U}_p$ and standard deviation σ_p, which are used to compute the dissipation rate D_p included in the wave energy or action equation.

This simple permeable layer model in the wet zone is coupled with the combined wave and current model above the permeable layer to predict irregular wave breaking and transmission over a submerged porous breakwater. The coupled model was compared with wave-flume experiments with no landward water flux ($q_o = 0$) by Kobayashi *et al.* (2007b) and Garcia and Kobayashi (2015). The degree of agreement (within errors of about 20%) is similar for impermeable beaches and permeable submerged structures.

The permeable layer model is extended to a permeable wet and dry zone above the still water level. A number of time-dependent hydrodynamic models for rubble mound structures were developed to predict the temporal and spatial variations of wave dynamics as accurately as possible. The computation time normally increases with the increase in the resolution and accuracy. The computationally advanced models are used to predict hydrodynamic variables for relatively short durations. To reduce computation time considerably, Kobayashi *et al.* (2007b) proposed the probabilistic model CSHORE. The time-varying wave variables are expressed using a probability distribution. The spatial variations of the mean and standard deviation are computed using the time-averaged governing equations. The probabilistic time-averaged model requires additional assumptions, but its computational efficiency allows the calibration of the model parameters using a large number of tests. This probabilistic model for the wet zone on the permeable armor layer was extended by Kobayashi *et al.* (2010a, 2011) to the wet and dry zone in order to predict wave runup and overtopping. The extended model provides the hydrodynamic input to a damage or erosion progression model that predicts the slow evolution of the stone or gravel layer profile.

The movement of individual stone units on the armor layer may be computed using the equation of motion for each armor unit (Kobayashi and Otta, 1987). The profile evolution of the armor layer may then be predicted by computing the displacements of all the armor units. However, this approach has never been adopted for practical applications probably because of its computation time and the uncertainty of initial individual stone locations. The sediment transport model in CSHORE is modified to predict the profile evolution of the stone or gravel layer in the same manner as the prediction of the sand beach profile evolution. This simple approach neglects the discrete nature of armor stone units but is very convenient for the prediction of the armor layer profile evolution averaged alongshore where the alongshore averaging reduces the discrete nature (Melby and Kobayashi, 1998).

The landward-marching computation using the model for the wet zone is continued as long as the computed $\bar{h}$ is larger than 0.1 cm. A separate model for the wet and dry zone is developed and connected with the model for the wet zone (Kobayashi *et al.*, 2010a). This procedure is the same as that used for the impermeable beach and structure. The time-averaged cross-shore continuity and momentum equations derived from the nonlinear shallow-water wave equations on the permeable slope (Wurjanto and Kobayashi, 1993) include the time-averaged vertical seepage velocity and horizontal momentum flux into the permeable layer. The continuity and momentum equations for the flow inside the permeable layer are used to predict the vertical seepage velocity and horizontal momentum flux in simplified manners. In short, the probabilistic swash model for the impermeable bottom has been extended to allow water volume and momentum fluxes between the flows above and inside the permeable layer.

Kobayashi *et al.* (2010a) compared the extended CSHORE with the 52 tests by van Gent (2002) and Kobayashi and de las Sautos (2007). The seaward slope was in the range of 1/5–1/2. The nominal stone diameter D_{n50} varied from 0.49 to 4.23 cm. The maximum vertical thickness of the armor layer was in the range of 0.49–14.0 cm. The measured porosity of the stone was $n_p = 0.5$. The agreement for the wave overtopping rate q_o was mostly within 100% errors.

The sediment transport model for the impermeable sand beach is modified to predict the deformation of gravel or stone structures on a fixed bottom. The stone is characterized by the nominal stone diameter D_{n50} and specific gravity s. The probability P_b of gravel or stone movement is estimated by assuming that the stone movement occurs when the instantaneous velocity U exceeds the critical velocity estimated using the critical stability number N_c for the stone, which is of the order of unity (Kobayashi *et al.*, 2003) and related to the critical Shields parameter for the initiation of sand movement. The probability of stone suspension is estimated using the stone fall velocity w_f but the computed probability of stone suspension is essentially zero. The stone armor units are assumed to move like bedload particles. For the computation for arbitrary stone or gavel, suspended load is included to allow the comparison of bed load and suspended load. The formulas for bed load and suspended load are assumed to be the same for sand, gravel, and stone in which the permeability and roughness differences are accounted for in the hydrodynamic models for the impermeable and permeable bottoms. CSHORE cannot predict the movement of sand and stone (or gravel) simultaneously.

Kobayashi *et al.* (2010a) compared the stone CSHORE with the three damage progression tests by Melby and Kobayashi (1998). The armor stone was placed in a traditional two-layer thickness with a seaward slope of 1/2. The armor stone was characterized by $D_{n50} = 3.64$ cm, $s = 2.66$, and $n_p = 0.4$. The thickness of the armor layer was 7.3 cm. The test duration was in the range of 8.5–28.5 h. The numerical model overpredicted the deposited area below SWL at the end of the test mostly because it does not account for discrete stone units dislodged and deposited at a distance seaward of the toe of the damaged armor layer. The eroded area above SWL was predicted better. The temporal variation of the eroded area A_e was compared using damage S_e defined as $S_e = A_e/D_{n50}^2$. The numerical model predicted the damage progression well partly because the critical stability number N_c was calibrated to be $N_c = 0.7$ for the three damage progression tests. The temporal variations of S_e computed for $N_c = 0.7$ and 0.6 were fairly sensitive to N_c.

Kobayashi *et al.* (2011) compared CSHORE with four gravel beach evolution tests conducted in a wave flume. The median gravel diameter was 2.0 mm and the fall velocity was 25 cm/s. The profile changes of two erosion tests on a steep slope of 1/2 were predicted well by CSHORE with $N_c = 0.7$. The accretional change on a mild slope of 1/5 and the onshore bar migration and formation of an equilibrium gravel beach profile were reproduced sufficiently after the bedload parameter b was increased with the increase in the fraction Q of irregular breaking waves. The bed load formula for onshore gravel and stone transport is still uncertain and will need to be improved using additional laboratory and field data.

Kobayashi *et al.* (2013c) extended the model for wave runup and overtopping to the landward wet zone and predicted the temporal variations of the damage and wave transmission of a reef breakwater during a storm. Intermittent plunging water produces complex hydrodynamics in the landward wet zone. The mean $\bar{\eta}$ and standard deviation σ_η for the free surface elevation η in the landward wet zone are simply assumed to be constant and the same as those computed at the landward end of the wet and dry zone. The mean $\bar{U}$ and standard deviation σ_U of the cross-shore velocity U are computed using the continuity equation and the linear wave relation between η and U. The mean $\bar{V}$ and standard deviation σ_V of the longshore velocity V are assumed to be negligible. The cross-shore and longshore stone transport rates in the landward wet zone are computed using the formulas used in the seaward wet zone. The extended model was compared with the 148 tests of Ahrens (1989) for a reef breakwater with a narrow crest at or above SWL in which the narrow crest was lowered by wave action. The numerical model simulates the transition from an emerged crest to a submerged crest. The model was also compared with the experiment of Ota *et al.* (2006) on a wide-crested submerged breakwater in which the crest height increased during 20-h wave action. The damage, crest height, and wave transmission coefficient were predicted reasonably well, but the damaged profile was not predicted accurately. The rudimentary model for the landward wet zone will need to be improved with the aid of detailed measurements in the zone of plunging water above the landward slope.

Garcia and Kobayashi (2015) applied CSHORE extended to the landward wet zone to examine the spatial variation of damage on different sections of the trunk and head of a low-crested stone structure under normally and obliquely incident irregular waves. The computed wave transmission coefficient and damage on the front slope, back slope, and total section of the trunk were compared with the 188 tests of Vidal and Mansard (1995) and Kramer and Burcharth (2003). The similarity of the trunk and head damage for the low-crested breakwater was proposed to predict damage on the front head and back head using the numerical model for the trunk. The agreement was mostly within a factor of 2 after the calibrations of $N_c = 0.6$. The numerical model may not be very accurate but provides an additional tool for the design of low-crested stone structures based on laboratory experiments (Burcharth *et al.*, 2006). A numerical wire mesh was created to mimic the wire mesh used by Vidal and Mansard (1995) who examined the stability of different stone sizes on the front slope, crest, and back slope of a low-crested rubble mound structure. This numerical wire mesh may be applied to design spatially varying stone sizes on a rubble mound structure.

The components of CSHORE from 2003 to 2013 were developed for common coastal engineering applications in the U.S., such as dune erosion and overwash; beach erosion and recovery; wave runup, overtopping, and transmission over coastal structures; and rubble mound damage. The CSHORE extensions from 2014 to 2024 were aimed at other specific applications of practical importance but limited knowledge. The extensions are separated into the following six themes: (1) woody plants and piles, (2) dike and clay erosion, (3) barrier beach overtopping and bay flooding, (4) sand and stone interactions, (5) intertidal mudflat, and (6) beach fill design. The CSHORE extension for the given theme is explained in such a way that coastal engineers and scientists may be able to judge the applicability of each CSHORE extension to their specific problems. The details of each CSHORE extension are available in the journal publications cited in the following. The research reports published by the Center for Applied Coastal Research (CACR), University of Delaware, are posted on the CACR webpage (www.coastal.udel.edu). These reports include those written by graduate students and visiting researchers who contributed to the development of CSHORE.

6. Pile Effects on Dune Erosion and Overwash

Vegetation may be present on natural beaches, and buildings may exist on urban beaches. A cluster of discrete objects modify waves and currents and affect beach profile changes during storms. Interactions of waves and aquatic vegetation (e.g., kelp) were investigated earlier (e.g., Kobayashi *et al.*, 1993). Kobayashi *et al.* (2013a) conducted five tests in a wave flume to examine the effects of wooden dowels (idealized tree stems) on erosion and overwash of high and low sand dunes. A wide vegetation covering the high dune reduced scarping prevented wave overtopping initially and reduced sand overwash after the initiation of wave overtopping. A wide vegetation zone covering an entire low dune reduced dune erosion by retarding wave uprush and reducing wave overtopping and overwash. Ayat and Kobayashi (2015) conducted additional four tests to examine the placement density and toppling effects of the dowels. The effectiveness of the dowels in reducing dune erosion and overwash decreased significantly with the density decrease and dowel toppling. Ayat and Kobayashi (2015) extended CSHORE to include the drag force acting on the dowel. The drag coefficient was calibrated within the range of 1–2 on the basis of available wave force data for a single pile (Tørum, 1989). The cross-shore variations of the mean and standard deviation of the free surface elevation and cross-shore velocity were predicted within errors of about 20%. The profile evolutions of the dunes with and without dowels were predicted by a factor of about 2. The effectiveness of the wide dowel zone covering the high and low dunes in reducing the wave overtopping and sand overwash rates was also reproduced, but these small rates are difficult to predict accurately. The details of these two studies were published in the CACR reports of CACR-12-05 and CACR-13-07.

Quan and Kobayashi (2015) examined the utility of a pile fence in reducing dune erosion and overwash during a storm. Six tests were conducted to compare the effectiveness of six different pile fences. The experiment suggested that the pile fence with a porosity of about 0.5 should be placed near the toe of the dune foreslope with a sufficient burial depth to avoid toppling. The fence porosity is defined as the fraction of alongshore opening encountered by uprush and downrush. The measured time series were used to propose the exceedance probability distributions of the free surface elevation and onshore velocity. The measured extreme values (5%, 1%, and 0.013% exceedance probabilities) were predicted by CSHORE with about 20% errors. The measured dune crest lowering in the presence of a pile fence was predicted fairly, but the steeply scarped foreslope was underpredicted. The details of this study were published in the report of CACR-14-09.

Shore protection projects require the prediction of coastal storm damage and economic loss. Cárdenas and Kobayashi (2017) conducted 11 tests on a sand beach to examine the movement of 10 wooden blocks (floatable objects) placed on the foreshore and berm as well as short and long pilings. The initial block elevation above the sand surface had little effect on the hydrodynamics, sediment transport, and profile evolution in this experiment with widely spaced blocks. The block floating (by buoyancy force) and sliding (by wave drag force) on the sand surface and the block falling from the pilings depended on the swash hydrodynamics and block clearance above the foreshore and berm whose profile varied during each test. A simple probabilistic model was developed to estimate the immersion, sliding, and floating probabilities for the blocks in the swash zone. The predicted probabilities are compared with the observed cross-shore variation of the block response on or above the accretional and erosional beach profiles which influenced the block response noticeably. The details of this study were published in the report of CACR-15-08.

7. Erosion of Grassed Dike and Clay

Large-scale laboratory and real dike experiments were performed to understand the erosion processes of grassed earthen dikes (levees) mostly in the Netherlands (Steendam *et al.*, 2010, 2012). Kobayashi and Weitzner (2015) developed a dike erosion model to predict the evolution of the eroded dike profile during a storm. The developed model was calibrated and verified using available data (Smith *et al.*, 1994; Wolters *et al.*, 2008). The resistance force of the turf per unit horizontal area is denoted as (ρR) where ρ is the fluid density and R is the resistance force divided by ρ so that its unit is $(m/s)^2$. The value of $\sqrt{R}$ is the velocity scale for the resistance force. The rate of erosion work is expressed as the product of the resistance force and the rate of the vertical erosion depth increase. The rate of erosion work is assumed to be provided by the wave energy dissipation rates per unit horizontal area due to wave breaking and bottom friction. For the dike turf, the resistance parameter R is assumed to decrease linearly from the turf surface to the underneath clay and remain constant in the clay layer. The turf thickness is typically about 0.1 m. The calibrated values of R based on full-scale testing were 1,000 and 200 $(m/s)^2$ for good (dense roots) and poor grass covers, respectively, and 10 $(m/s)^2$ for boulder clay with a network of cracks formed under long-term weathering (Kobayashi and Weitzner, 2014). The detail of this study was published in the report of CACR-14-01.

Kobayashi and Zhu (2020) formulated the erosion processes of consolidated cohesive sediment (glacial till) under irregular breaking waves to predict the profile evolution of a cohesive sediment beach with a layer of sand (USACE, 2003). The dike erosion model with no turf above the clay layer is used to predict the erosion of consolidated cohesive sediment containing cohesionless sediment (sand). Sand released from the eroded sediment is transported onshore or offshore by wave action (Nairn and Southgate, 1993). The cohesive sediment erosion is increased by the abrasive effect of a thin mobile sand layer and decreased by the protective effect of a thick sand layer. The sand layer evolution on the impermeable cohesive sediment is predicted using the sand transport model in CSHORE including limited sand availability on the eroding cohesive sediment. This extended CSHORE was compared with the flume experiment data by Skafel (1995) who used till blocks excavated from a site on the north shore of Lake Erie. The measured till erosion rates were on the order of 0.05 cm/h. The calibrated R for the glacial till was 30 (m/s)^2 in comparison to 10 (m/s)^2 for the boulder clay with a network of cracks. The detail including applications was published in the report of CACR-19-01.

Cliff erosion is a visible factor for shoreline retreat in California (Lester *et al.*, 2022). Zhu and Kobayashi (2021a) investigated soft cliff (bluff) erosion using the wave basin data by Damgaard and Dong (2004) and the cross-shore model CSHORE for dune erosion (Kobayashi *et al.*, 2009, 2010b). The measured cliff recession rates under oblique breaking waves for cliffs built of wet sand (no clay) in 15 tests could be reproduced by CSHORE with a single cross-shore line and a new option that accounts for sand loss associated with the alongshore gradient of longshore sand transport in Eq. (1) using an equivalent alongshore distance (dimensional calibration parameter). The effect of sediment cohesion on cliff erosion was examined using CSHORE extended to cohesive sediment containing sand (Kobayashi and Zhu, 2020). Damgaard and Dong (2004) included one test with a sand/clay mixture (9.2% clay). The calibrated resistance parameter was $R_c = 1$ (m/s)^2 and the computed cliff recession rate for $R_c < 1$ (m/s)^2 was limited by the removed rate of sand deposited at the toe of the eroding bluff, as shown in Fig. 2. The detail of this study was published in the report of CACR-19-05. Leone *et al.* (2022) applied CSHORE to examine the measured erosion of a consolidated sand dune using a mineral colloidal silica-based grout.

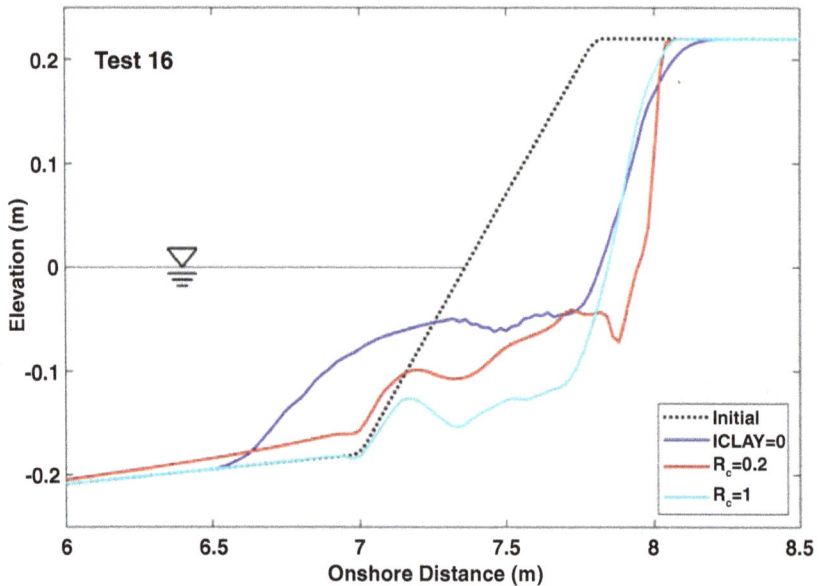

Fig. 2.   Initial (time $t = 0$) and computed sand profiles z_b at time $t = 3$ h for Test 16 (Damgaard and Dong, 2004) with ICLAY=0 (no clay) and ICLAY=1 (sand/clay mixture and erosion resistance parameter $R_c = 0.2$ and 1 m²/s²).

8. Wave Overtopping of Barrier Beach and Bay Flooding

Low-laying barrier beaches and islands are common along the U.S. East Coast and the Gulf of Mexico. Wave overtopping of a low-laying sand barrier may increase the water level in an enclosed bay or lagoon. CSHORE was used to predict wave overtopping and overwash of dunes (Figlus *et al.*, 2011; Johnson *et al.*, 2012) but the destination of overtopped water was not computed. Kobayashi *et al.* (2013c) extended CSHORE to the landward wet zone and computed wave transmission over and through a porous breakwater. For an emerged sand barrier, the still water levels in the ocean and bay may be different and wave overtopping can increase the bay water level. Kobayashi and Zhu (2017) computed wave overtopping and overwash over the barrier beach between the Atlantic Ocean and the Rehoboth Bay in Delaware during Hurricane Sandy in 2012 where the measured still water level in the bay was specified as the landward boundary condition. The barrier beach was eroded but remained above the ocean

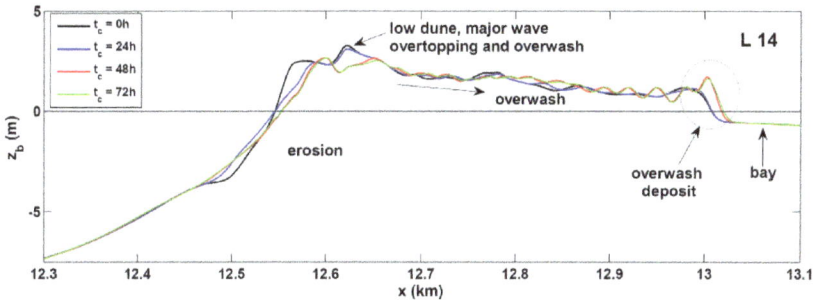

Fig. 3. Initial ($t_c = 0$) and computed beach profile (onshore distance $x = 0$ in 20 m water depth) at computation time $t_c = 24$, 48, and 72 h along cross-shore line L14 for Hurricane Sandy where time t was used for the bay water level analysis by Kobayashi and Zhu (2017).

and bay still water levels. The computed wave overtopping rate per unit width on the order of 0.1 m²/s (100 l/s/m) may have increased the peak bay water level by 0.1–0.2 m (10–20%). The detail of this study including an analytical model for the peak bay water level was published in the report of CACR-17-01. Melby *et al.* (2021) conducted a coastal Texas flood risk assessment using a Monte Carlo life-cycle simulation approach with CSHORE embedded in the modeling system in order to estimate the frequency of overflow and rehabilitation.

9. Sand Transport on and Inside Porous Stone Structures

Low rock structures have been constructed on some beaches to reduce storm-induced damage to backshore areas (Irish *et al.*, 2013), but no design guideline is available to design such structures. Kobayashi and Kim (2017) conducted four tests in a wave flume with a sand beach and a berm to compare the effectiveness of a narrow dune and a rock (stone) seawall placed on the foreshore in reducing wave overtopping and sand overwash. The dune was effective in eliminating or reducing wave over-topping compared with the corresponding berm with no dune, but the narrow dune was destroyed easily as the water level was increased. The stone seawall was effective in reducing wave overtopping and overwash even after it was deformed by stone settlement. A stone seawall buried inside a dune functioned like the dune initially and like the seawall after the sand on and inside the porous seawall was eroded by overtopping waves. CSHORE was extended to predict sand transport on and inside the porous structure. The fabric filter under the rock seawall was assumed to be fixed because sand transport processes under the filter were unknown. The simple extension of CSHORE was based on the conservation of sand volume in the pores of the stone structure as well as an empirical reduction factor for sand mobility in the pores. The extended CSHORE predicted little sand deposition inside the exposed seawall and extensive sand removal from the buried seawall. However, the agreement with the profile data in the vicinity of the seawall was marginal partly because of the assumption of no settlement of the seawall. The details of this study were published in the report of CACR-17-02.

Design manuals for stone structures against waves (e.g., USACE, 2003) do not account for sand and stone interactions. Yuksel and Kobayashi (2020) conducted three tests in a wave flume to compare sand beach profile evolution and wave overtopping of a sand berm for the cases of no structure, a stone revetment protecting the steep sand berm, and a stone sill reducing wave action on the berm. The same stones were used to construct the revetment and sill. The revetment reduced onshore sand transport on the sand beach in front of the revetment and was effective in protecting the sand berm and reducing wave overtopping. The settlement of the revetment placed on a filter occurred, and the revetment crest was damaged during major wave overtopping. The sill reduced the beach profile change but was not very effective in reducing wave overtopping and berm erosion when the sill crest was submerged sufficiently. Piled stones on the narrow sill crest were displaced. The lowered and wider crest became stable. The settlement of the sill placed on the filter did not occur perhaps because the sill was placed on the stable section of the beach in the test of no structure. CSHORE was upgraded for its application to the sill test where the emerged sill crest became submerged during the test. The reduced beach profile changes in the presence of the revetment and sill were reproduced, but CSHORE could not produce sufficient onshore sand transport near the shoreline. The upgraded CSHORE included an option of no filter to predict the settlement of the stone structure caused by sand erosion below the structure and estimate the settlement reduction provided by the filter. The filter in the revetment test may have reduced the revetment settlement by about 0.1 m. For the sill test of no filter settlement, the settlement reduction provided by the filter was computed to be small. The upgraded CSHORE may be used to estimate the degree of the stone structure settlement for the case of no filter and the beach profile change in the vicinity of the stone structure placed on the filter. The details of this study were published in the report of CACR-19-04.

A rubble mound structure was constructed to close a breached channel through a barrier island (Webb *et al.*, 2011). This emergency project provided an example of sand and stone interactions. Zhu and Kobayashi (2021b) investigated irregular wave overwash and landward migration of a narrow sand barrier in a wave flume. The emerged crest of a sand barrier was lowered rapidly to the still water level (SWL) and migrated landward slowly with its crest slightly below the SWL. A rock mound consisting of three layers of stable stones was constructed on the submerged sand barrier without any filter. The rock mound settled but its crest remained above the SWL. The landward migration of the sand barrier was reduced only partially because of onshore sand transport over and through the porous structure. The sand barrier was rebuilt with a rock cover on the crest of the initial sand barrier. The rock cover consisting of a single layer of stable stone reduced the sand barrier deformation only slightly because the stone settlement and spreading exposed underlying sand to direct wave action. CSHORE was compared with the three tests (no rock, rock mound, and rock cover). The mean and standard deviation of the free surface elevation and cross-shore velocity were predicted within errors of 20% as in previous comparisons except that small transmitted waves were difficult to predict accurately. The barrier profile evolution was predicted marginally because of the difficulty in predicting the cross-shore variation of the sand transport rate over the barrier crest and through the porous structure. The details of this study were published in the report of CACR-21-01.

Beach recovery occurred seaward of the rubble mound structure constructed in 2011 to close a 2 km wide breach through Dauphin Island created by Hurricane Katrina in 2005 (Gongalez *et al.*, 2020). Zhu and Kobayashi (2023) analyzed the shoreline positions of the closed Katrina Cut from 2015 to 2020 using the topography of the five aerial images in Fig. 4. The dry beach width increased by more than 80 m seaward of the Katrina Cut structure and decreased up to 37 m on the neighboring beach. The onshore sand transport rate computed using CSHORE was large enough to cause the observed beach recovery starting from the measured bathymetry in 2015. The recovering beach reduced the depth-limited breaking wave height and wave action on the rubble mound structure

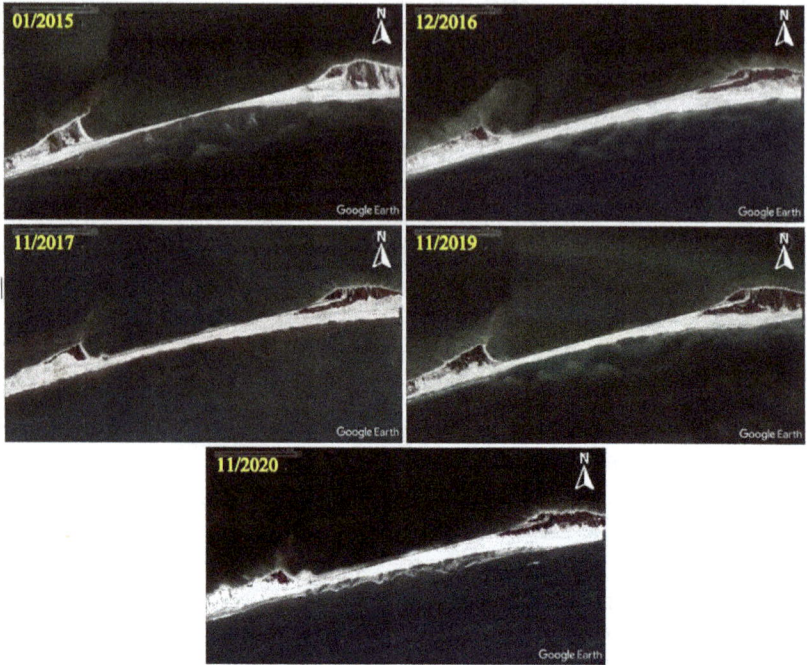

Fig. 4. Google Earth photos (2015–2020) of Katrina Cut in Dauphin Island, Alabama, between the Gulf of Mexico (below, south) and the Mississippi Sound (above, north).

during Hurricane Nate. This temporary structure induced the breached beach recovery and may have become more lasting thanks to the increasing protection ensured by the front beach. The details of this study were published in the report of CACR-22-02.

The rehabilitation of damaged or aged rubble mound structures has not been studied in a systematic manner. Melby *et al.* (2015) performed a risk assessment of damaged breakwaters in Rhode Island. The rehabilitation timing of a damaged rubble mound structure depends on the remaining capacity of the deteriorated structure. Strazzella and Kobayashi (2022) conducted a laboratory experiment in a wave flume to measure the remaining capacity of a low-crested rubble mound inside the surf zone on a sand beach. The formation of a bar and trough feature modified wave conditions at the toe of the structure. Wave transmission over and through the structure increased with the lowering of the mound crest. The mound with a double armor layer on smaller core stones was exposed to irregular wave action lasting 22.2 hours. Some of the core stone was visible through holes in the thinned armor layer but remained in place. Crest lowering reduced the wave action on the damaged mound. The mound with a single armor layer did not stabilize itself because the core stone was removed after 7.8 hours. Strazzella and Kobayashi (2024) extended CSHORE to predict the damage progression of armor and core stones. The extended CSHORE reproduced the measured damage progression after the calibration of the armor stone critical stability number and the core stone resistance parameter. The details of the experiment were published in the report of CACR-22-03.

10. Intertidal Mudflat Profile Evolution

For typical prototype CSHORE computations, tide effects are included in the input time series of the still water level at the seaward boundary located in a water depth of about 10 m. For a macrotidal beach with a gentle bottom slope, the large temporal change of water volume in the computation domain produces measurable cross-shore tidal currents. Do *et al.* (2016) analyzed 17-day field data in Korea to investigate the cross-shore wave transformation, currents, and sand suspension in the intertidal zone, and beach profile changes of a macrotidal beach with a gentle slope of 0.02 and tidal ranges of 4–7 m. The measured beach profile changes were less than 0.5 m despite the occurrence of two consecutive storms with significant wave heights exceeding 2 m. CSHORE including cross-shore tidal current was shown to reproduce the gradual wave height decay in the migratory surf zone, the cross-shore current of the order of 0.1 m/s affected by flood and ebb tides, the suspended sand volume varying with the tidal cycle, and the small beach profile changes. CSHORE can predict the morphology of the macrotidal sand beach.

Mudflat morphology is poorly understood due to complex cohesive sediment dynamics. The mudflat is normally composed of a mixture of sand and mud. Miranda and Kobayashi (2022) extended CSHORE to the sand and mud mixture and predicted the erosional and accretional mudflat profile changes measured almost monthly by Yamada and Kobayashi (2004) from 03 July 2001 to 08 August 2002. The field site was located on the coast of Ariake Bay in Japan. The cross-shore survey line was marked by 30 wooden stakes driven into the mudflat at an interval of 50 m. The mudflat elevation could always be measured for the survey points located between 100 and 1,050 m from the seawall near the mouth of the Shirakawa River. The average significant wave height and period were

0.2 m and 3 s, respectively, because of the limited fetch lengths in this closed shallow bay. The tide is semidiurnal and the average tidal range was 2.88 m. The moving averaged water level revealed the yearly oscillation of 0.4 m in height where this water level was higher in summer and lower in winter. The sediment characteristics within 2 m below the mudflat surface were fairly uniform perhaps because of bioturbation. The sand mass, mud mass including organic matter, and water mass per unit volume of the mixed sediment were approximately 400 kg/m^3, 400 kg/m^3, and 500 kg/m^3, respectively. The sediment characteristics are assumed to remain constant during the computation interval. The sand and mud interactions are neglected. The sediment transport model in CSHORE is used to predict the cross-shore and longshore sand transport rates. Mud is assumed to be transported as a suspended load. The mud transport rate is estimated by modifying the formula for suspended sand transport in CSHORE.

For the comparison of the extended CSHORE with the intertidal mudflat data, the semidiurnal migration of the still water shoreline and surf zone is resolved numerically to predict the net cross-shore and longshore sediment transport rates influenced by the small cross-shore (undertow) and longshore currents induced by breaking waves of about 0.2 m height. Alongshore sediment loss or gain is included by approximating the alongshore sediment transport gradient using an equivalent alongshore length in Eq. (1). The calibrated CSHORE reproduces the measured erosional (accretional) profile change of about 0.1 m (0.1 m) over the cross-shore distance of 950 m during the erosional (accretional) interval of 206 (195) days. The mudflat profile changes are equally affected by the mud characteristics, the semidiurnal tide amplitude, and the wave height, period, and direction. In addition, the alongshore water level gradient and wind stress influence longshore current and sediment transport. This study shows the importance of sediment transport in the surf zone which may have been excluded in previous numerical modeling of intertidal mudflat profile evolution. The details of the CSHORE extension, computation, and comparison were published in the report of CACR-22-01.

11. Beach Fill Design Under Low Wave Energy

Beach fill design guidelines in the United States (USACE, 2003) were based mostly on the experiences of beach nourishment projects along open coastlines exposed to hurricanes and winter storms. Figlus and

Kobayashi (2008) analyzed a beach nourishment project along the Delaware Atlantic beaches consisting of medium sands under high wave energy. The fill volume per unit alongshore length was in the range of 144–250 m³/m in 1998. The beaches were renourished in 2005 or 2007 (7 or 9 years later). The first major beach nourishment in Thailand was carried out at Pattaya Beach which is microtidal and under low wave energy (Laksanalamai and Kobayashi, 2022). Pattaya Beach was widened by placing about 130 m³/m of medium sand ($d_{50} = 0.35$ mm) along the shoreline length of 2.8 km between two terminal groins constructed in 2018. The bathymetry and topography were measured in 2015 before the sand placement and yearly from 2019 to 2023. The beach profile and sand volume changes were analyzed by Laksanalamai and Kobayashi (2024). The relatively small (less than 10%) sand volume changes from 2019 to 2023 suggested stabilization from disturbances created by the nourishment project. The nourishment project and data analyses were published in the reports of CACR-21-02 and CACR-24-01.

The nourished Pattaya Beach is depicted in Fig. 5 where the terminal north (south) groin of 60 m length interrupts northward (southward) sand transport. The Google Earth Pro image of October 22, 2022, was taken almost four years after the sand placement.

Laksanalamai and Kobayashi (2024) proposed a novel analytical model based on an equilibrium foreshore slope. The model predicts offshore (onshore) sand transport on the foreshore slope which is steeper (milder) than the equilibrium foreshore slope. A laboratory experiment was conducted to verify the model concept. The verified model was applied to decipher the apparent stability of the nourished beach and to estimate the degree of alongshore sand loss or gain. This diagnostic model may eventually be incorporated into short-term foreshore profile forecasting.

Beach fill design may be improved if the destinations of sand particles placed on eroding beaches can be predicted. Laksanalamai and Kobayashi (2023) conducted an experiment in a wave flume to measure the trajectories of 20 small objects (gravel and microplastics) in the surf and swash zones on an equilibrium beach and a nourished foreshore beach. Gravel particles were mobile within the swash zone. CSHORE was extended to track small submerged discrete objects using the hydrodynamic and sediment transport variables computed in CSHORE. The simple tracking model calibrated for the gravel and microplastics could predict limited net displacements of hypothetical sand particles on the equilibrium sand beach.

Fig. 5.   Photos of 2019 nourished beach with the north groin (top left) and the south groin (bottom left) and Pattaya satellite image (Google Earth Pro) on October 22, 2022.

Laksanalamai and Kobayashi (2025) extended the tracking model to predict both cross-shore and longshore trajectories of small discrete objects. The extended CSHORE was applied to elucidate the apparent stability of the nourished foreshore and berm from 2019 to 2023 of no surge and wave data. The computations were limited to the duration of one day with specified tide (1.5 m tidal range) or surge (1.79 m maximum) and significant wave heights of 0.2 m, 1 m, and 2 m, corresponding to normal, monthly large, and 20-year extreme waves. The nourished 2019 profile is computed to be slightly (0.006 m) accretional under the tide and 0.2 m waves and erosional (0.2 m) under the tide and 1 m waves. The computed trajectories of the medium sand particles are onshore except for the sand particles in the surf zone under the 1 m waves. For the surge and 2 m waves, the 2019 profile is eroded up to 0.4 m and the computed wave overtopping rate of the seawall is of the order of a tolerable rate with possible flooding landward of the wall. The tracked sand particles converge at the toe of the foreshore for the medium sand of the median diameter $d_{50} = 0.35$ mm but the fine sand of $d_{50} = 0.18$ mm could be transported farther offshore and deposited in deeper water. The sand size of the beach fill will affect its longevity on the foreshore.

12. Conclusions

The combined wave–current model CSHORE based on the time-averaged continuity, cross-shore momentum, longshore momentum, wave action or energy, and roller energy equations predicts the cross-shore variations of the mean and standard deviation of the free surface elevation and depth-averaged cross-shore and longshore velocities under normally or obliquely incident irregular breaking waves. The sediment transport formulas for the cross-shore and longshore transport rates of suspended sediment and bedload are relatively simple and require the hydrodynamic input variables which can be predicted efficiently and fairly accurately using existing wave and current models. The simple model CSHORE has been compared with a larger number of small-scale and large-scale laboratory data and field data. CSHORE has been extended to the intermittently wet and dry zone for the prediction of wave overwash of dunes and deformation of low-crested stone structures and gravel beaches.

The improvement of CSHORE requires the simultaneous measurements of hydrodynamics and sediment dynamics because of the interactions among waves, sediment, and bottom elevation changes in the surf and swash zones. The simplicity and computational efficiency of CSHORE should allow its application to various coastal engineering problems that tend to occur in the vicinity of the shoreline. The development of CSHORE up to 2013 was focused on common coastal engineering applications in the U.S. such as dune erosion and overwash; beach erosion and recovery; wave runup, overtopping, and transmission over coastal structures; and damage to rubble mound breakwaters. CSHORE 2013 was limited to cohesionless sediments (sand, gravel, and stone).

The CSHORE extensions from 2014 to 2024 were aimed at other specific applications such as the use of woody plants and piles to reduce dune erosion, the erosion of grassed dike and consolidated cohesive sediment, wave overtopping and overwash of a barrier beach, sand transport on and inside a porous stone structure, intertidal mudflat profile evolution, and beach fill design under low wave energy. Additional empirical formulas and parameters were introduced for these extensions and calibrated using small-scale laboratory experiments conducted for specific CSHORE extensions or available large-scale laboratory data and scarce field data.

The next stage of CSHORE evolution may be the utilization of CSHORE in devising innovative solutions for coastal engineering problems. Coastal engineering problems are international but affordable solutions depend on local conditions. Consequently, we need a range of solutions. In addition, the predictive accuracy of CSHORE should be improved. The capabilities and shortcomings of CSHORE and XBEACH (Roelvink *et al.*, 2009) were evaluated in the comparison with beach profile evolution data during storms in southern California (Kalligeris *et al.*, 2020) and in Duck, North Carolina (Cohn *et al.*, 2021). Both models were not very accurate.

Data Availability Statement

The manual and open source code of CSHORE 2024 (no pre and post processing routines) are available at the following site managed by the second author, Tingting Zhu:

https://drive.google.com/drive/folders/1srAM9BvnN7DX3qrsvVns1yId 1tLCCaiL?usp=sharing.

Acknowledgments

The following graduate students and researchers contributed to the development of the numerical model CSHORE: Brad Johnson, Yuki Tega, Lizbeth Meigs, Haoyu Zhao, Andres Payo, Paco de los Santos, Arpit Agarwal, Jeff Melby, Ali Farhadzadeh, Mark Gravens, Jens Figlus, Lauren Schmied, Mitch Buck, Elizabeth Hicks, Jill Pietropaolo, Hooyoung Jung, Kideok Do, Christine Grahler, Take Saitoh, Berna Ayat, Heather Weitzner, Roland Garcia, Rebecca Quan, Xavier Chaves Cardenas, Tugce Yuksel, Hirokazu Sumi, Elisa Leone, Tingting Zhu, Hyun Dong Kim, Jirat Laksanalamai, Sravani Mallavarapu, Jowi Miranda, and Michele Strazzella.

References

Ahrens, J. P. (1989). Stability of reef breakwaters. *Journal of Waterway, Port, Coastal, Ocean Engineering*, 115(2), 221–234.

Ayat, B. and Kobayashi, N. (2015). Vertical cylinder density and toppling effects on dune erosion and overwash. *Journal of Waterway, Port, Coastal, Ocean Engineering*, 141(1), 04014026.

Bagnold, R. A. (1966). *An Approach to the Sediment Transport Problem from General Physics.* U.S. Geological Survey, Prof. Paper 422-I.

Battjes, J. A. and Stive, M. J. F. (1985). Calibration and verification of a dissipation model for random breaking waves. *Journal of Geophysical Research: Oceans*, 90(C5), 9159–9167.

Burcharth, H. F., Kramer, M., Lamberti, A., and Zanuttigh, B. (2006). Structural stability of detached low crested breakwaters. *Coastal Engineering Journal*, 53(4), 381–394.

Cárdenas, C. X. and Kobayashi, N. (2017). Cross-shore damage variation of wooden blocks in swash zone on sand beach. *Coastal Engineering Journal*, 59(1), 175001-1.

Cohn, N., Brodie, K. L., Johnson, B., and Palmsten, M. L. (2021). Hotspot dune erosion on an intermediate beach. *Coastal Engineering Journal*, 170, 103998.

Cox, D. T. and Kobayashi, N. (2000). Identification of intense, intermittent coherent motions under shoaling and breaking waves. *Journal of Geophysical Research: Oceans*, 105(C6), 14223–14236.

Cox, D. T., Kobayashi, N., and Okayasu, A. (1996). Bottom shear stress in the surf zone. *Journal of Geophysical Research: Oceans*, 101(C6), 14337–14348.

Dalrymple, R. A. (1988). Model for refraction of water waves. *Journal of Waterway, Port, Coastal, and Ocean Engineering*, 114(4), 423–435.

Damgaard, J. S. and Dong, P. (2004). Soft cliff recession under oblique waves: Physical model tests. *Journal of Waterway, Port, Coastal, and Ocean Engineering*, 130(5), 234–242.

Dean, R. G. (1991). Equilibrium beach profile: Characteristics and applications. *Journal of Coastal Research*, 53–84.

Do, K., Kobayashi, N., Suh, K.-D. and Jin, J.-Y. (2016). Wave transformation and sand transport on a macrotidal pocket beach. *Journal of Waterway, Port, Coastal, and Ocean Engineering*, 142(1), 04015009.

Dohmen-Janssen, C. M. and Hanes, D. H. (2002). Sheet flow dynamics under monochromatic nonbreaking waves. *Journal of Geophysical Research: Oceans*, 107(C10), 13-1–13-21.

Dohmen-Janssen, C. M., Kroekenstoel, D. F., Hassan, W. N., and Ribberink, J. S. (2002). Phase lags in oscillatory sheet flow: Experiments and bed load modeling. *Coastal Engineering*, 47, 295–327.

EurOtop Manual (2007). *Wave Overtopping of Sea Defenses and Related Structures: Assessment Manual.* www.overtopping-manual.com.

EurOtop Manual (2018). *Manual on Wave Overtopping of Sea Defenses and Related Structures.* www.overtopping-manual.com.

Farhadzadeh, A., Kobayashi, N., and Gravens, M. B. (2012). Effect of breaking waves and external current on longshore sediment transport. *Journal of Waterway, Port, Coastal, and Ocean Engineering*, 138(3), 256–260.

Figlus, J. and Kobayashi, N. (2008). Inverse estimation of sand transport rates on nourished Delaware beaches. *Journal of Waterway, Port, Coastal, and Ocean Engineering*, 134(4), 218–225.

Figlus, J., Kobayashi, N., and Gralher, C. (2012). Onshore migration of emerged ridge and ponded runnel. *Journal of Waterway, Port, Coastal, and Ocean Engineering*, 138(5), 331–338.

Figlus, J., Kobayashi, N., Gralher, C., and Iranzo, V. (2011). Wave overtopping and overwash of dunes. *Journal of Waterway, Port, Coastal, and Ocean Engineering*, 137(1), 26–33.

Garcia, R. and Kobayashi, N. (2015). Trunk and head damage on a low crested breakwater. *Journal of Waterway, Port, Coastal, and Ocean Engineering*, 141(2), 04014037.

Gonzalez, V. M., Garcia-Moreno, F. A., Melby, J. A., Nadal-Caraballo, N. C., and Godsey, E. S. (2020). Alabama Barrier Island Restoration Assessment life-cycle structure response modeling. *Rep. ERDC/CHL TR-20-5*. Vicksburg, MS: US Army Engineer Research and Development Center.

Hoagland, S. W., Jeffries, C. R., Irish, J. L., Weiss, R., Mandli, K., Vitousek, S., Johnson, C. M., and Cialone, M. A. (2023). Advances in morphodynamic modeling of coastal barriers: A review. *Journal of Waterway, Port, Coastal, and Ocean Engineering*, 149(5), 03123001.

Holland, K. T., Holman, R. A., and Sallenger, A. H., Jr. (1991). Estimation of overwash bore velocities using video techniques. *Proceedings of Coastal Sediments '91*, ASCE, Reston, VA, 489–497.

Irias Mata, M. and van Gent, M. R. A. (2023). Numerical modeling of wave overtopping discharges at rubble mound breakwaters using OPENFOAM®. *Coastal Engineering*, 181, 104274.

Irish, J. L., Lynett, P. J., Weiss, R., Smallegan, S. M., and Cheng, W. (2013). Buried relic seawall mitigates Hurricane Sandy's impacts. *Coastal Engineering*, 80, 79–82.

Johnson, B. D., Kobayashi, N., and Gravens, M. B. (2012). Cross-shore numerical model CSHORE for waves, currents, sediment transport and beach profile evolution. *Final Rep. No. ERDC/CHL TR-12-22*, U.S. Army Corps of Engineers, Coastal and Hydraulics Laboratory, Vicksburg, MS.

Kalligeris, N., Smit, P. B., Ludka, B. C., Guza, R. T. and Gallien, T. W. (2020). Calibration and assessment of process-based numerical models for beach profile evolution in southern California. *Coastal Engineering*, 158, 103650.

Kamphuis, J. S. (1991). Alongshore sediment transport rate. *Journal of Waterway, Port, Coastal, and Ocean Engineering*, 117(6), 624–640.

Kobayashi, N. (2015). Hydraulic response and armor layer stability on coastal structures. *Res. Rep. No. CACR-15-07*, Center for Applied Coastal Research, University of Delaware, Newark, DE.

Kobayashi, N. (2016). Coastal sediment transport modeling for engineering applications. *Journal of Waterway, Port, Coastal, and Ocean Engineering*, 142(6), 03116001.

Kobayashi, N. and Johnson, B. D. (2001). Sand suspension, storage, advection, and settling in surf and swash zones. *Journal of Geophysical Research: Oceans*, 106(C5), 9363–9376.

Kobayashi, N. and de los Santos, F. J. (2007). Irregular wave seepage and over-topping of permeable slopes. *Journal of Waterway, Port, Coastal, and Ocean Engineering*, 133(4), 245–254.

Kobayashi, N. and Jung, H. (2012). Beach erosion and recovery. *Journal of Waterway, Port, Coastal, and Ocean Engineering*, 138(6), 473–483.

Kobayashi, N. and Kim, H. D. (2017). Rock seawall in the swash zone to reduce wave overtopping and overwash of a sand beach. *Journal of Waterway, Port, Coastal, and Ocean Engineering*, 143(6), 04017033.

Kobayashi, N. and Otta, A. K. (1987). Hydraulic stability analysis of armor units. *Journal of Waterway, Port, Coastal, and Ocean Engineering*, 113(2), 171–186.

Kobayashi, N. and Tega, Y. (2002). Sand suspension and transport on equilibrium beach. *Journal of Waterway, Port, Coastal, and Ocean Engineering*, 128(6), 234–248.

Kobayashi, N. and Weitzner, H. (2015). Erosion of a seaward dike slope by wave action. *Journal of Waterway, Port, Coastal, and Ocean Engineering*, 141(2), 04014034.

Kobayashi, N. and Wurjanto, A. (1990). Numerical model for waves on rough permeable slopes. *Journal of Coastal Research*, SI(7), 149–166.

Kobayashi, N. and Wurjanto, A. (1992). Irregular wave setup and runup on beaches. *Journal of Waterway, Port, Coastal, and Ocean Engineering*, 118(4), 368–386.

Kobayashi, N. and Zhu, T. (2017). Bay flooding through tidal inlet and by wave overtopping of barrier beach. *Journal of Waterway, Port, Coastal, and Ocean Engineering*, 143(5), 04017024.

Kobayashi, N. and Zhu, T. (2020). Erosion by wave action of consolidated cohesive bottom containing cohesionless sediment. *Journal of Waterway, Port, Coastal, and Ocean Engineering*, 146(2), 04019041.

Kobayashi, N. and Zhu, T. (2022). Cross-shore numerical model CSHORE 2022 for coastal sediments and structures. *Res. Rep. No. CACR-22-04*, Center for Applied Coastal Research, University of Delaware, Newark, Delaware.

Kobayashi, N., Agarwal, A., and Johnson, B. D. (2007a). Longshore current and sediment transport on beaches. *Journal of Waterway, Port, Coastal, and Ocean Engineering*, 133(4), 296–304.

Kobayashi, N., Buck, M., Payo, A., and Johnson, B. D. (2009). Berm and dune erosion during a storm. *Journal of Waterway, Port, Coastal, and Ocean Engineering*, 135(1), 1–10.

Kobayashi, N., de los Santos, F. J., and Kearney, P. G. (2008a). Time-averaged probabilistic model for irregular wave runup on permeable slopes. *Journal of Waterway, Port, Coastal, and Ocean Engineering*, 134(2), 88–96.

Kobayashi, N., DeSilva, G. S., and Watson, K. D. (1989). Wave transformation and swash oscillation on gentle and steep slopes. *Journal of Geophysical Research: Oceans*, 94(C1), 951–966.

Kobayashi, N., Farhadzadeh, A., and Melby, J. A. (2010a). Wave overtopping and damage progression of stone armor layer. *Journal of Waterway, Port, Coastal, and Ocean Engineering*, 136(5), 257–265.

Kobayashi, N., Farhadzadeh, A., Melby, J., Johnson, B., and Gravens, M. (2010b). Wave overtopping of levees and overwash of dunes. *Journal of Coastal Research*, 26(5), 888–900.

Kobayashi, N., Gralher, C., and Do, K. (2013a). Effects of woody plants on dune erosion and overwash. *Journal of Waterway, Port, Coastal, and Ocean Engineering*, 139(6), 466–472.

Kobayashi, N., Herrman, M. N., Johnson, B. D., and Orzech, M. D. (1998). Probability distribution of surface elevation in surf and swash zones. *Journal of Waterway, Port, Coastal, and Ocean Engineering*, 124(3), 99–107.

Kobayashi, N., Hicks, B. S., and Figlus, J. (2011). Evolution of gravel beach profiles. *Journal of Waterway, Port, Coastal, and Ocean Engineering*, 137(5), 258–262.

Kobayashi, N., Meigs, L. E., Ota, T., and Melby, J. A. (2007b). Irregular breaking wave transmission over submerged porous breakwater. *Journal of Waterway, Port, Coastal, and Ocean Engineering*, 133(2), 104–116.

Kobayashi, N., Otta, A. K., and Ray, I. (1987). Wave reflection and runup on rough slopes. *Journal of Waterway, Port, Coastal, and Ocean Engineering*, 113(3), 282–298.

Kobayashi, N., Payo, A., and Schmied, L. (2008b). Cross-shore suspended sand and bedload transport on beaches. *Journal of Geophysical Research: Oceans*, 113(C7).

Kobayashi, N., Pietropaolo, J. A., and Melby, J. A. (2013b). Wave transformation and runup on dikes and gentle slopes. *Journal of Coastal Research*, 29(3), 615–623.

Kobayashi, N., Pietropaolo, J. A., and Melby, J. A. (2013c). Deformation of reef breakwaters and wave transmission. *Journal of Waterway, Port, Coastal, and Ocean Engineering*, 139(4), 336–340.

Kobayashi, N., Pozueta, B., and Melby, J. A. (2003). Performance of coastal structures against sequences of hurricanes. *Journal of Waterway, Port, Coastal, and Ocean Engineering*, 129(5), 219–228.

Kobayashi, N., Raichle, A., and Asano, T. (1993). Wave attenuation by vegetation. *Journal of Waterway, Port, Coastal, and Ocean Engineering*, 119(1), 30–48.

Kobayashi, N., Zhao, H., and Tega, Y. (2005). Suspended sand transport in surf zones. *Journal of Geophysical Research: Oceans*, 110(C12).

Kobayashi, N., Zhu, T., and Mallavarapu S. (2018). Equilibrium beach profile with net cross-shore sand transport. *Journal of Waterway, Port, Coastal, and Ocean Engineering*, 144(6), 04018016.

Kramer, M. and Burcharth, H. (2003). DELOS: Wave basin experiment final form: 3D stability tests at AAU. *Rep. EVK-CT-2000-0004*, Hydraulics and Coastal Engineering Group, Dept. of Civil Engineering, Aalborg University, Aalborg, Denmark.

Laksanalamai, J. and Kobayashi, N. (2022). Performance of a nourished sand beach in the upper Gulf of Thailand. *Journal of Coastal and Offshore Science and Engineering*, 1(2), 46–54.

Laksanalamai, J. and Kobayashi, N. (2023). Tracking of small discrete objects submerged in surf and swash zones on sand beaches. *Journal of Waterway, Port, Coastal, and Ocean Engineering*, 149(6), 04023017.

Laksanalamai, J. and Kobayashi, N. (2024). Equilibrium foreshore slope and cross-shore sediment transport rate. *Journal of Coastal Research* (Submitted)

Laksanalamai, J. and Kobayashi, N. (2025). Performance evaluation of a beach nourishment project under low wave energy in Thailand. *Coastal Engineering Journal* (submitted).

Leone, E., Kobayashi, N., D'Alessandro, F., and Tomasicchio, G. R. (2022). Prediction of consolidated sand dune erosion by waves. *Journal of Coastal and Offshore Science and Engineering*, 1(2), 56–70.

Lester, C., Griggs, G., Patsch, K., and Anderson, R. (2022). Shoreline retreat in California: Taking a step back. *Journal of Coastal Research*, 38(6), 1207–1230.

Losada, I. J., Lara, J. L., Guanche, R., and Gonzalez-Ondina, J. M. (2008). Numerical analysis of wave overtopping of rubble mound breakwaters. *Coastal Engineering Journal*, 55, 47–62.

Madsen, O. S. and Grant, W. D. (1976). Quantitative description of sediment transport by waves. *Proceedings of the 15th Coastal Engineering Conference*, ASCE, Reston, VA, 1093–1112.

Madsen, O. S., Chisholm, T. A., and Wright, L. D. (1994). Suspended sediment transport in inner shelf waters during extreme storms. *Procerence 24th Coastal Engineering Confeedings of*, ASCE, Reston, VA, 1849–1864.

Mase, H. (1989). Random wave runup height on gentle slope. *Journal of Waterway, Port, Coastal, and Ocean Engineering*, 115(5), 649–661.

Masselink, G., Austin, M. J., O'Hare, T. J., and Russell, P. E. (2007). Geometry and dynamics of wave ripples in the nearshore zone of a coarse sandy beach. *Journal of Geophysical Research: Oceans*, 112(C10).

Mei, C. C. (1989). *The Applied Dynamics of Ocean Surface Waves*. World Scientific, Singapore.

Melby, J. A. and Kobayashi, N. (1998). Progression and variability of damage on rubble mound breakwaters. *Journal of Waterway, Port, Coastal, and Ocean Engineering*, 124(6), 286–294.

Melby, J. A. and Kobayashi, N. (2011). Stone armor damage initiation and progression based on maximum momentum flux. *Journal of Coastal Research*, 27(1), 110–119.

Melby, J. A., Massey, T. C., Das, H. S., Nadal-Caraballo, N. C., Gonzalez, V. M., Bryant, M. A., Tritinger, A. S., Provost, L. A., Owensby, M. B., Stehno, A. L. and Diop, F. (2021). Coastal Texas Protection and Restoration Feasibility Study. *Rep. ERDC/CHL TR-21-11*. Vicksburg, MS: US Army Engineer Research Development Center.

Melby, J. A., Nadal-Caraballo, N. C., and Kobayashi, N. (2012). Wave runup prediction for flood mapping. *Proceedings of 32nd Coastal Engineering Conference*, Management 79, 1–15.

Melby, J. A., Nadal-Caraballo, N. C., and Winkelman, J. H. (2015). Point Judith, Rhode Island, Breakwater Risk Assessment. *Rep. ERDC/CHL TR-15-13*. Vicksburg, MS: US Army Engineer Research Development Center.

Miranda, P. S. and Kobayashi, N. (2022). Numerical modeling of intertidal mudflat evolution under waves and currents. *Coastal Engineering Journal*, 64(3), 406–427.

Nairn, R. B. and Southgate, H. N. (1993). Deterministic profile modeling of nearshore processes. Part 2. Sediment transport and beach profile development. *Coastal Engineering Journal*, 19, 57–96.

Ota, T., Kobayashi, N., and Kimura, A. (2006). Irregular wave transformation over deforming submerged breakwater. *Proceedings of the 30th Coastal Engineering Conference*, World Scientific, Singapore, 4945–4956.

Payo, A., Kobayashi, N., and Yamada, F. (2009). Suspended sand transport along pier depression. *Journal of Waterway, Port, Coastal, and Ocean Engineering*, 135(5), 245–249.

Phillips, O. M. (1977). *The Dynamics of the Upper Ocean*. Cambridge University Press, Cambridge, U.K.

Quan, R. and Kobayashi, N. (2015). Pile fence to reduce wave overtopping and overwash of dune. *Journal of Waterway, Port, Coastal, and Ocean Engineering*, 141(6), 04015005.

Ribberink, J. S. (1998). Bedload transport for steady flow and unsteady oscillatory flows. *Coastal Engineering Journal*, 34, 59–82.

Ribberink, J. S. and Al-Salem, A. A. (1994). Sediment transport in oscillatory boundary layers in cases of rippled beds and sheet flow. *Journal of Geophysical Research: Oceans*, 99(C6), 12707–12727.

Roelvink, D., Reniers, A., Van Dongeren, A. P., De Vries, J. V. T., McCall, R., and Lescinski, J. (2009). Modelling storm impacts on beaches, dunes and barrier islands. *Coastal Engineering Journal*, 56(11-12), 1133–1152.

Saitoh, T. and Kobayashi, N. (2012). Wave transformation and crossshore sediment transport on sloping beach in front of vertical wall. *Journal of Coastal Research*, 28(2), 354–359.

Saunders, T. M., Cohn, N., and Hesser, T. (2024). Insights into nearshore sandbar dynamics through process-based numerical and logistic regression modeling. *Coastal Engineering Journal*, 192, 104558.

Schüttrumpf, H. and Oumeraci, H. (2005). Layer thickness and velocities of wave overtopping flow at seadikes. *Coastal Engineering Journal*, 52, 473–495.

Skafel, M. G. 1995. Laboratory measurement of nearshore velocities and erosion of cohesive sediment (till) shorelines. *Coastal Engineering Journal*, 24(3-4), 343–349.

Smith, G. M., Seijffert, J. W. W., and Van der Meer, J. W. (1994). Erosion and overtopping of a grass dike: Large scale model tests. *Coastal Engineering Proceedings*, 24(1994), 2639–2652.

Steendam, G. J., Provoost, Y., and van der Meer, J. (2012). Destructive wave overtopping and wave runup tests on grass covered slopes of real dikes. *Coastal Engineering Proceedings*, 33(2012), 1–14.

Steendam, G. J., van der Meer, J., Hardeman, B., and van Hoven, A. (2010). Destructive wave overtopping tests on grass covered landward slopes of dikes and transitions to berms. *Coastal Engineering Proceedings*, 32(2010), 1–14.

Strazzella, M. and Kobayashi, N. (2022). Remaining capacity of low-crested rubble mounds damaged by breaking waves in surf zone. *Journal of Waterway, Port, Coastal, and Ocean Engineering*, 148(6), 04022020.

Strazzella, M. and Kobayashi, N., (2024). Damage progression of low-crested rubble mound with single layer and core. In *Coasts, Marine Structures and Breakwaters 2023: Resilience and Adaptability in a Changing Climate* (pp. 1215–1219). Emerald Publishing Limited.

Svendsen, I. A., Haas, K. and Zhao, Q. (2002). Quasi-3D nearshore circulation model SHORECIRC version 2.0. *Res. Rep. No. CACR-02-01*, Center for Applied Coastal Research, University of Delaware, Newark, Delaware.

Tørum, A. (1989). Wave forces on pile in surface zone. *Journal of Waterway, Port, Coastal, and Ocean Engineering*, 115(4), 547–565.

Trowbridge, J. and Young, D. (1989). Sand transport by unbroken water waves under sheet flow conditions. *Journal of Geophysical Research: Oceans*, 94(C8), 10971–10991.

USACE (U.S. Army Corps of Engineers). (2003). *Coastal Engineering Manual*. Washington, D.C.

van Gent, M. R. A. (1995). Porous flow through rubble-mound material. *Journal of Waterway, Port, Coastal, and Ocean Engineering*, 121(3), 176–181.

van Gent, M. R. A. (2001). Wave runup on dikes with shallow foreshores. *Journal of Waterway, Port, Coastal, and Ocean Engineering*, 127(5), 254–262.

van Gent, M. R. A. (2002). Low-exceedance wave overtopping events: Measurements of velocities and the thickness of water-layers on the crest and inner slope of dikes. *Delft Cluster Report DC030202/H3803*, Delft Hydraulics, Delft, The Netherlands.

Vidal, C. and Mansard, E. P. D. (1995). On the stability of reef breakwaters. *Tech. Rep.*, National Research Council of Canada, Ottawa.

Webb, B. M., Douglass, S. L., Dixon, C. R., and Buhring, B. (2011). Coast guards. *Civil Engineering Magazine Archive*, 81(12), 76–83.

Wolters, G., Nieuwenhuis, J. W., van der Meer, J., and Klein Breteler, M. (2008). Large scale tests of boulder clay erosion at the Wieringermeer dike (Ijsselmeer). *Proceedings of the 31st Coastal Engineering Conference*, World Scientific, Singapore, 3263–3275.

Wurjanto, A. and Kobayashi, N. (1993). Irregular wave reflection and runup on permeable slopes. *Journal of Waterway, Port, Coastal, and Ocean Engineering*, 119(5), 537–557.

Yamada, F. and Kobayashi, N. (2004). Annual variations of tide level and mudflat profile. *Journal of Waterway, Port, Coastal, and Ocean Engineering*, 130(3), 119–126.

Yuksel, Z. T. and Kobayashi, N. (2020). Comparison of revetment and sill in reducing shore erosion and wave overtopping. *Journal of Waterway, Port, Coastal, and Ocean Engineering*, 146(1), 04019028.

Zhu, T. and Kobayashi, N. (2021a). Modeling of soft cliff erosion by oblique breaking waves during a storm. *Journal of Waterway, Port, Coastal, and Ocean Engineering*, 147(4), 04021009.

Zhu, T. and Kobayashi, N. (2021b). Rock mound to reduce wave overwash and crest lowering of a sand barrier. *Coastal Engineering Journal*, 63(3), 1–13.

Zhu, T. and Kobayashi, N. (2023). Recovery of eroded beach seaward of closed Katrina Cut in Dauphin Island, Alabama. *Journal of Coastal and Offshore Science and Engineering*, 2(1), 1–12.

Chapter 4

Global Climate Change and Resilient Coastal Structures

Yalcin Yuksel

*Department of Civil Engineering, Yildiz Technical University,
Davutpasa Campus, 34220 Esenler, Istanbul, Turkey*

yalcinyksl@gmail.com

The vulnerability of coastal areas and marine ecosystems to climate change is increasing. Rising sea levels, more frequent and intense storms, ocean acidification, and other climate-related factors pose significant threats to coastal areas and coastal structures. To protect coastal structures and marine habitats, it is essential to develop effective adaptation and mitigation strategies. Performance-based design emerges as a crucial approach for ensuring the durability of coastal structures. To reduce the risks associated with rising sea levels and intensifying storms, we must upgrade existing coastal structures or design new ones that are more resistant. Many existing coastal structures, such as breakwaters, may not be sufficient to counter the increasing risks posed by climate change. These structures may need to be upgraded, modified, or reinforced with additional elements. Innovative design approaches are necessary to improve coastal structures and ensure their resilience against the negative effects of global climate change. Additionally, building with nature is a comprehensive engineering approach that leverages natural ecological processes to achieve sustainable coastal structure designs. It offers a promising alternative to conventional hard coastal defenses, often leading to negative environmental impacts. Nature-based solutions can enhance coastal resilience, improve water quality, support biodiversity, and promote carbon sequestration.

1. Climate Change and Extreme Events

The vulnerability of coastal areas and marine ecosystems has increased with the increase in human pressures and climate change impacts. This situation creates the following problems in the coastal regions: (i) social and economic loss, (ii) property and infrastructure damage, (iii) coastal waters and erosion problem, (iv) risk to human health and safety, and (v) endangerment of coastal wetlands and biological diversity and marine ecosystems.

Global climate change has serious impacts on the ecology and morphology of coastal and marine areas due to rising atmospheric carbon dioxide, variations in the hydrologic cycle, ocean temperature and acidity, increasing temperatures, sea-level rise, increased ocean stratification, decreased sea-ice extent, and altered patterns of ocean circulation, precipitation, and freshwater input. As a result, coastal and marine resilience is an important environmental adaptation and mitigation strategy.

Coastal areas are projected to have more permanent inundation and flooding threats from sea-level rise, intense rains, high tide flooding, and severe storms in the future due to climate change. Scientists project that as the climate warms, there will be more intense hurricanes as well as increased rainfall (IPCC, 2021). Coastal storms are expected to be the most destructive climate-related hazard, causing coastal flooding and coastal erosion (Bergillos *et al.*, 2019; Sardella *et al.*, 2020; Rodriguez-Delgado *et al.*, 2020). Other climate threats to coasts include ocean acidification, harmful algal blooms, and saltwater intrusion.

The reports provided by the Intergovernmental Panel on Climate Change (IPCC, Sixth Assessment Report, AR6) (2023) and the UN Environment Program (UNEP) (2023) address the impacts of climate change on coastal regions and oceans, highlighting the profound and multifaceted challenges facing marine ecosystems and the urgent need for coordinated action to protect ocean health. These reports detail how climate change affects the oceans through events such as ocean warming, acidification, sea-level rise, and changes in ocean circulation patterns which have far-reaching impacts on marine biodiversity, ecosystems, and the millions of people who depend on the oceans for their livelihoods and food security. The reports indicate a continued rise in sea levels, primarily caused by the thermal expansion of seawater and the melting of glaciers, which make the coastal communities increasingly vulnerable to flooding, erosion, and saltwater intrusion, posing risks to infrastructure, economies, and ecosystems. Increasing greenhouse gas emissions lead to more heat absorption by the oceans, causing ocean temperatures to rise. The absorption of excess carbon dioxide in the atmosphere by the oceans causes seawater to become more acidic. Ocean acidification has been noted to threaten the health of marine biodiversity, fisheries, and coral reef ecosystems, with potential cascading effects on coastal economies and livelihoods.

Changing climate will lead to a change in wind characteristics (Hdidouan and Staffell, 2017) and then to a change in oceanic response (Chowdhury *et al.*, 2019). Coastal regions face the increasing frequency and intensity of extreme weather events, such as storms, hurricanes, and coastal flooding exacerbated by climate change. Therefore, there is a widespread interest in determining the best estimate of climate data that requires realistic models for the past, present, and future. In recent years, global and regional models have been widely used to infer future projections at global and regional scales (Hemer *et al.*, 2013; Lemos *et al.*, 2019; Meucci *et al.*, 2020; Soares *et al.*, 2017; Alvarez and Lorenzo, 2019; Wang *et al.*, 2020; Islek *et al.*, 2022a, 2022b).

One of the most significant impacts of global climate change is the intensification and prolongation of extreme storms. The most recent report from the Intergovernmental Panel on Climate Change (IPCC, 2023) emphasized that projected changes in the frequency and intensity of extreme events are becoming more pronounced, underscoring the importance of studying these events across various climate components. Additionally, the IPCC (2023) reported that global surface temperatures (including both land and ocean) have risen since 1901, primarily due to increased concentrations of greenhouse gases (GHGs) resulting from human activities. This warming has accelerated the rate of global mean sea-level rise since the early 1900s (IPCC, 2021). Understanding the spatial and temporal variability of mean sea-level pressure (MSLP) and the occurrence of its extremes is crucial for comprehending the complexities of the climate system. A known relationship exists where a reduction in atmospheric pressure by 1 hPa (or 1 mbar) causes a ~1 cm rise in sea level, known as the inverted barometer effect (Brown *et al.*, 1975). Recent research by Islek and Yuksel (2023) indicates that significant warming in sea surface temperature (SST) has led to a notable decrease in MSLP and an increase in sea level in the eastern Mediterranean and Black Seas in recent years. Climate change is expected to alter the spatial and temporal characteristics of wind (including intensity, frequency, direction, duration, and extreme events), which will, in turn, affect Earth's weather systems. In the context of climate change, these effects are particularly evident in extreme wave heights.

Global climate change poses greater risks in the design of coastal and offshore structures than wind and wave climate predictions made using historical data (Yuksel *et al.*, 2020). As a result, it is important to determine design wave parameters using long-term wave data, with datasets spanning at least 30 years. As an example, reliable design wave heights were determined by analyzing wave data modeled with EC-EARTH regional wind scenarios data (Strandberg *et al.*, 2014), covering both historical (1970–2005) and future (2021–2100) periods. For this purpose, IPCC AR5 used regional climate scenarios (RCPs) for future projections. These RCPs (RCP2.6, RCP4.5, RCP6.0, and RCP8.5) are based on radiative forcing values ranging from 2.6 W/m² to 8.5 W/m². Emissions under RCP2.6 are expected to peak between 2010 and 2020, while RCP4.5 is expected to peak around 2040, RCP6.0 is expected to peak around 2080, and RCP8.5 is expected to increase continuously and rapidly throughout the 21st century (Meinshausen *et al.*, 2011). As a result of this example, extreme value distributions for the Aegean Sea and the Eastern Mediterranean are presented in Fig. 1, respectively. Extreme wave statistical analysis was performed using the GEV (Generalized Extreme Value) distribution, which combines the Gumbel, Fréchet, and Weibull distributions into a single expression. Figure 1 shows that, both in the near future (2021–2060) and medium future (2061–2100), the extreme value curves for the RCP4.5 scenario are seen to be larger than in the historical period (1979–2005). The RCP4.5 scenario predicted extreme values for both the Aegean Sea and the Eastern Mediterranean compared to the historical period, both in the near (2021–2060) and in the middle (2061–2100) future. The projected increase in significant wave height for the 100-year return period will reach 12% in the Aegean Sea in the near future, while an increase of 17% will occur in the medium future. In the Eastern Mediterranean, the increase in both the near and medium future is 5.85% (Gumuscu, 2024).

As shown in Fig. 1, relying solely on historical data to estimate future extreme significant wave heights may lead to underestimations due to the

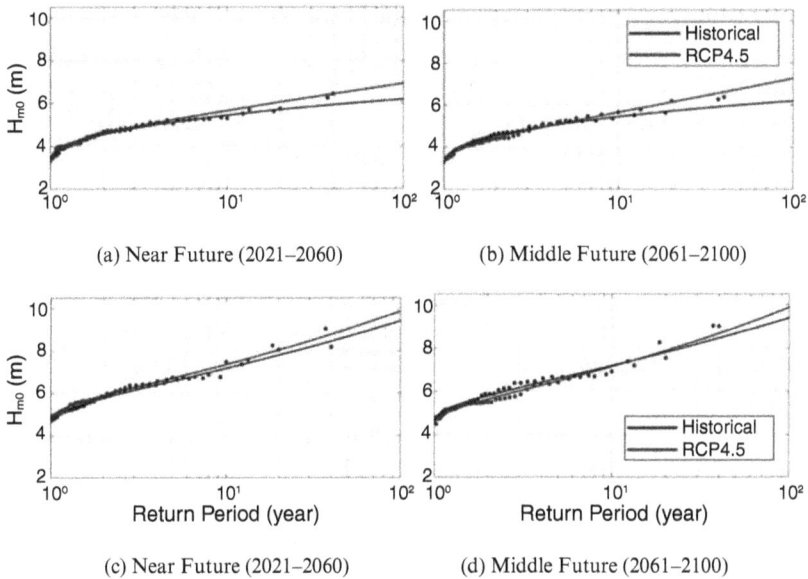

(a) Near Future (2021–2060)   (b) Middle Future (2061–2100)

(c) Near Future (2021–2060)   (d) Middle Future (2061–2100)

Fig. 1.   Significant wave height based on annual occurrence probabilities for the near future and middle future periods in Argean Sea (a, b) and Eastern Mediterranean (c, d) where black and blue lines shows historical and RCP4.5 extremes, respectively (revised from Gumuscu, 2024).

impacts of climate change. Therefore, analyses should account for regional climate change to improve accuracy.

Coastal environments are dynamic and complex systems shaped by various factors, with wave climate being a key driver of change. Waves, as significant factors of geomorphic change, play a central role in shaping shorelines and influencing sediment transport along the coast. Coastal regions, critical interfaces between land and sea, are experiencing new transformations driven by the wide-ranging effects of climate change. Shoreline evolution is considerably impacted by changes in climate patterns, rising sea levels, and extreme weather events, creating complicated challenges for coastal areas worldwide (Ranasinghe *et al.*, 2012, 2013; Zacharioudaki and Reeve, 2011; Ashton *et al.*, 2008; Vitousek *et al.*, 2017).

Adaptation and mitigation efforts are crucial to address the impacts of climate change on the oceans. Adaptation measures include coastal protection, sustainable fisheries management, sustainable coastal development, and ecosystem-based approaches to marine conservation. Mitigation efforts include reducing greenhouse gas emissions to limit further warming and sea-level rise through measures, such as switching to renewable energy sources, promoting energy efficiency, and reducing deforestation. These adaptation and mitigation measures require international cooperation and policy responses to address interconnected challenges such as climate change and ocean conservation, which includes implementing Paris Agreement targets to limit global warming, developing marine protected areas, reducing pollution and overfishing, and supporting vulnerable coastal communities to adapt to the effects of climate change (UNEP, 2023). IPCC (2023) also highlighted the urgent need for coordinated action at local, national, and global levels to protect and sustainably manage coastal zones and oceans in the face of ongoing climate change.

All these studies agree on the need to implement nature-compatible solutions in coastal areas and oceans. Failure to address these challenges risks irreversible damage to marine ecosystems, coastal communities, and the global economy. However, it is still unclear what these solutions will be and how they will be implemented. Research on the ocean–climate interface will strengthen the global scientific capacity to better understand the drivers of change in the ocean, to address emerging threats, to provide solutions and innovative approaches, and to support decision-making in the areas of climate change mitigation and adaptation, and the maintenance of a healthy ocean and marine environment.

In Section 2, how to determine performance-based design parameters due to global climate change for coastal structures is first explained. In Section 3, how upgrading of breakwaters and coastal revetments, which are coastal protection structures, and how nature-based solutions can be designed against the effects of global climate change are discussed.

2. Basic Principles of Performance-Based Design for Coastal Structures

Performance-based design emerged in the 1990s (PIANC, 2001) as a response to lessons learned from earthquakes. Later, similar concepts began to be used for the performance of port and coastal structures under wave effect (Yuksel *et al.*, 2020). Its goal is to address the shortcomings of conventional design methods. Unlike conventional design, which focuses solely on providing sufficient capacity to withstand a design wave force, performance-based design considers the overall performance of a structure, including its ability to withstand forces beyond the design limit. This approach avoids the pitfalls of uneconomical designs for high return period waves and unsafe designs for low return period waves. Performance-based design establishes clear criteria for the importance of structures, their performance levels, and design foundations. It employs a systematic process that outlines specific performance requirements and corresponding evaluation methods.

Performance-based design criteria for coastal structures address various limit states to ensure safety, functionality, and resilience. These criteria are crucial for adapting to climate change, covering the increasing frequency and intensity of storms, hurricanes, and typhoons. By adopting performance-based design, coastal structures can better withstand the heightened loads and impacts associated with these extreme weather events.

Furthermore, this approach considers the environmental impact, contributing to the protection and preservation of local ecosystems. This is particularly important as climate change intensifies environmental stresses. Designing with resilience in mind promotes sustainable coastal development by minimizing future interventions and maintaining ecological balance.

Performance-based design also anticipates potential future changes, such as sea-level rise and more severe storms, ensuring that structures are adaptable to evolving conditions. This proactive approach enhances long-term resilience by allowing for future improvements and modifications. By adhering to principles outlined in British Standards and Eurocodes, performance-based design criteria are fundamental for creating coastal structures that are effective, safe, and sustainable in the face of climate change. These are as follows:

- **Ultimate Limit State (ULS):** Ultimate Limit State ensures the structure can withstand extreme conditions without collapsing or failing. The design should account for the maximum expected wave heights and storm surge levels based on historical data and future projections.
- **Serviceability Limit State (SLS):** Serviceability Limit State ensures the structure performs effectively under normal service conditions, without significant distress or functional issues. It should remain functional and usable both during and after typical loading conditions, including normal weather and sea conditions.

- **Durability:** Durability ensures the structure remains in good condition over its intended lifespan with minimal routine maintenance.
- **Adaptability and Resilience:** Adaptability and Resilience ensures the structure can adapt to future changes and recover from extreme events.
- **Economic Viability:** Economic Viability balances the cost of construction with long-term operational and maintenance costs.
- **Environmental Impact:** Environmental Impact minimizes adverse effects on the coastal and marine environment.

Performance-based design is comprehensively defined by Yuksel *et al.* (2018, 2020). This approach necessitates a reliability analysis, and probabilistic methods are particularly well suited for coastal facilities due to the irregular nature and fluctuating actions of waves. However, solely considering the probability of failure is inadequate when deformation (damage level) also needs to be factored in (Takahashi *et al.*, 2015; Yuksel *et al.*, 2016, 2018). Therefore, it is important to accurately define appropriate design wave levels and corresponding acceptable levels of structural damage.

By employing performance-based design, more precise approaches can be developed to address the extreme wave climates anticipated due to global climate change and the subsequent responses of coastal structures.

Several studies have explored the application of performance-based design to coastal structures (Ling *et al.*, 1999; PIANC, 2001; Goda, 2004; PIANC, 2014; Do *et al.*, 2016). While existing definitions for coastal structures under wave loads are not entirely sufficient, there is a clear need to establish design criteria for port and coastal protection structures that consider various design wave levels throughout their lifespans and incorporate probabilistic aspects.

Future changes in wave conditions due to climate change will significantly impact the marine ecosystem, coastal erosion, design of coastal defenses, performance of coastal and port structures, and overall coastal zone management. These changes will occur within the service life of existing coastal structures. Therefore, design wave parameters must be determined by considering projected wave conditions and sea-level rise. Performance-based design requires the establishment of extreme wave height distributions near the design site, which is the offshore boundary of a wave transformation model.

The new definitions for the functional classification of port and coastal structures and the definition of performance-damage levels, multi-level design wave levels, and performance objectives are briefly described with respect to the essentials of the performance-based design.

Performance-based design parameters:
1. Structural classes associated with the expected performance, usage, and functional importance,
2. performance levels associated with expected damage levels,
3. damage levels associated with frequent, rare, and very rare wave events,
4. performance objectives under different wave return period levels.

Functional classification of structures:
Coastal and port structures are classified as special, normal, simple, and unimportant structures:

(a) **Special Structures:** Structures to be used for rapid response and evacuation immediately after damage, and structures to be used for the marine structures of nuclear power plants, toxic, flammable, or explosive materials. Permanent changing of the nature such as inland dredging projects such as navigation straits.
(b) **Normal Structures:** Structures where the loss of life and property must be avoided, structures of economic and social significance, structures with difficult and time-consuming post-wave actions repair and retrofit needs, and port structures with crane operations.
(c) **Simple Structures:** Less important structures other than those classified in Special and Normal Structures, structures other than those classified as Unimportant Structures, port structures without crane operations such as only berthing usage of fishery ports.
(d) **Unimportant Structures:** Easily replaceable structures, structures not causing life safety risk even extensively damaged such as recreational coastal structures and sunbathing timber piers and temporary structures.

Performance levels:
Performance levels of coastal and port structures are defined with respect to expected damages during a storm wave event.

(a) **Minimum Damage (MD):** This performance level corresponds to a state where no or a very limited damage occurs in coastal and port structures and/or in their elements under a design wave and beyond event. In this case, port operation continues uninterruptedly or if any, service interruptions are limited to a few days.

(b) **Controlled Damage (CD):** This performance level corresponds to a state where non-extensive, repairable damage occurs in port structures and/or in their elements under a design wave. In this case, short-term (few weeks or months) interruptions in related port operations may be expected.

(c) **Extensive Damage (ED):** This performance level corresponds to a state where extensive damage occurs in coastal and port structures and/or in their elements under a design wave. In this case, long-term interruptions or even closures in related port operations may be expected.

(d) **State of Collapse (CS):** This corresponds to the collapse state in port structures and/or in their elements under the over design wave.

Design wave levels:
Four different levels of storms are defined in terms of their intensity, representing very frequent, frequent, rare, and very rare events:

(a) **Design Wave Level 1 (W1):** This design wave level represents very frequent and low-intensity design wave conditions with a high probability of occurring during the service life of port structures. The return period of (W1) design wave level is in between 10 and 50 years.

(b) **Design Wave Level 2 (W2):** This design wave level represents relatively frequent but low-intensity design wave conditions with a high probability of occurring during the service life of port structures. The return period of (W2) design wave level is in between 50 and 100 years.

(c) **Design Wave Level 3 (W3):** This design wave level represents the infrequent and high-intensity design wave conditions with a low probability of occurring during the service life of port structures. The return period of (W3) design wave level is 100 years.

(d) **Design Wave Level 4 (W4):** This design wave level represents the highest intensity and very infrequent design wave conditions. The return period of (W4) design wave level is more than 100 years with a 70% upper limit of the confidence interval for wind waves but 10,000 years for a tsunami wave.

The probability of occurrence of the design wave conditions (R, risk, %) during the service life of the structure should be determined at the design stage according to the service life of the structure (L) and design return period for design conditions (T_d):

$$R = 1 - (1 - 1/T_d)^L. \tag{1}$$

The acceptable risk levels for different T_d and L values are given in Table 1.

Occurrence probability is identified as an encounter probability of the design wave height during the service life of a structure. The return period can be determined by using the occurrence probability related to the service life of a structure. Hence, the design wave height can be obtained using the return periods defined in Table 2.

Since failures are not considered in the current design process, marine structure engineers do not pay sufficient attention to the extent and consequences of failure, i.e., performance evaluation is limited to the time prior to failure, while that during and after failure is neglected (Takahashi, *et al.*, 2015). However, the damage levels for coastal structures are classified according to their performance in Table 3.

Performance criteria (damage limits) such as displacement, tilting, settlement, and slope failure must be known for the coastal structures. Failure modes of typical coastal structures are shown in the Coastal Engineering Manual (CERC, 2003). CEM defines the failure in which

Table 1. Probability of occurrence of the design event R (%) of a coastal structure depending on a lifetime L (years) and design recurrence period Td (years) (PIANC, 2014).

Service Life,	Return Period, T_d (year)							
L (year)	5	10	30	50	100	500	1000	10 000
1	20	10	3	2	1	0	0	0
5	67	41	16	10	5	1	0	0
10	89	65	29	18	10	2	1	0
30	100	96	64	45	26	6	3	0
50	100	99	82	64	39	10	5	0
100	100	100	97	87	63	18	10	1
200	100	100	100	98	87	33	18	2
500	100	100	100	100	99	63	39	5

Table 2.   Return periods of wind-wave conditions for structural design (Yuksel *et al.*, 2020).

Structural Classes	Return Period (Year)	Structure Service Life (year)
Unimportant	10–50	1–20
Simple	50–100	30–70
Normal	100	50–100
Special	≥100(**)	200–beyond (*)

Notes: (*) Beyond case should be considered for nuclear power plants
(**) Return periods should be calculated using 70% upper limit of confidence interval for wind waves and return period is 10,000 years for tsunami.

Table 3.   Performance levels (Yuksel *et al.*, 2020).

Structural Classes	Damage Levels
Unimportant	ED or CD
Simple	CD or MD
Normal	MD
Special*	MD above the level of design wave condition

Note: *Failure conditions should be defined for breakwater design.

damage results in structural performance and functionality below the minimum anticipated by design. Damage is the partial collapse of a structure:

(i) Damage levels for breakwaters

Van der Meer (1988) defined the value of damage level of rubble mound breakwaters related to their slopes. Damage level is defined with $S = A_e/D_{n50}^2$ for rock units, where A_e is the area of displaced rocks/stones in the cross-section of the armor layer (including pores) above and below the design water level and D_{n50} is the nominal stone diameter. However, the damage level does not include the functional performance of structural classes. In performance evaluation of the rubble mound breakwaters for the case of ½ slope, the minimum damage (MD), controlled damage (CD), extensive damage (ED), and state of collapse (CS) may be considered as $S < 2$ (or no damage), $S = 2$, $S = 4$, and $S = 6$, respectively. This classification corresponds to the definitions in Table 3. Thus, the damage level of rubble mound breakwaters can be defined with respect to the functional performance of structural classes. As an example, if a rubble mound is a port structure, it must be defined as a normal structure, while a rubble mound is a special structure if its functional requirement is protecting a nuclear power plant.

Failure was defined in terms of exceedance of serviceability or ultimate limit states in Rock Manual (2007). Rock Manual (2007) indicated that the Serviceability Limit State (SLS) refers to the performance of the structure under normal conditions and generally defines the function the structure is required to perform: For example, a breakwater may be required to provide a certain level of protection to limit wave conditions in a harbor to an acceptable level. While exceedance of the SLS may not lead to the damage or failure of the breakwater, it will mean it is not performing the required function. However, the Ultimate Limit State (ULS) refers to performance under extreme conditions and generally defines the ability of the structure to survive under extreme loading conditions. Exceedance of the ULS leads to damage, and potentially failure, of the structure: For example, exceedance of design wave conditions may lead to the damage of breakwater armor and underlayers, with a risk of progressive failure (Rock Manual, 2007).

The damage mode of the monolithic vertical breakwaters is identified as horizontal displacement, settlement, and tilting. These modes are characterized by using quantitative damage levels and the damage levels of the monolithic vertical breakwater are summarized according to the performance-based design concept in Table 4. In Table 4, x is the sliding distance and H shows the height of the structure. If the designers choose the force-balance-based design, the safety factors are only used for the determination of the structure stability. When the performance-based design is considered, it is needed to determine the damage levels of the structures given in Table 4. Moreover, performance-based design can be performed by using advanced analysis, and these performance levels should be validated (calibrated) by physical model tests.

Goda and Takahashi (2001) described performance and reliability design and risk analysis as a new design methodology. Takahashi *et al.* (2000) concluded that the primary cause of damage is the sliding of the caisson. Any degree of sliding is equated with caisson damage. In practice, however, even if the caisson slides, the breakwater can still be functional, unless the sliding distance impedes the serviceability of the breakwater. The performance-based design method allows a certain

Table 4. Damage levels for caisson-type breakwater (Yuksel *et al.*, 2020).

	Damage Levels	
Performance Levels	Relative sliding (x/H) %	Tilting (°)
CS	Design engineer will define	
ED	5–7	5–10
CD	1.5–5	3–5
MD	<1.5	<3

Table 5. Allowable exceedance probability (%) of allowable sliding distance (OCDI, 2009).

	Structural Classes (cm)		
Damage Levels/(year)	Normal	Simple	Unimportant
CD (10)	15	30	50
ED (30)	5	10	20
CS (100)	2.5	5	10

amount of sliding during the lifetime of a breakwater (Suh *et al.*, 2012). Takahashi *et al.* (2000) defined the allowable expected sliding distance based on structural-functional classes:

- 3 cm for normal structure,
- 30 cm for simple structure,
- 100 cm for unimportant structure.

Moreover, Overseas Coastal Area Development Institute of Japan (2009) defined allowable exceedance probability (%) of allowable sliding distance (cm) and their structural classes in Table 5 where definitions of the classification correspond to the definitions in Table 3.

Breakwaters are not expected to be damaged under design wave condition; however, there is always a risk that a structure may be exposed to above the level of design wave height condition, and these conditions have not been taken into account at the design stage so far. Therefore, the failure performance of the structure under a higher level of the design wave height condition should be designated especially for special structures.

Numerical simulations have an important role in investigating wave transformations and wave actions on structures including wave forces, especially by the introduction of direct simulation techniques (Isobe *et al.*, 1999). Such simulations can explicitly show the process of wave propagation and action, which makes them quite suitable for performance design and use in the design process. Obviously then, both physical model experiments and numerical simulations are important tools in performance design (Takahashi *et al.*, 2015).

(ii) Effect of wave climate on performance-based design

In the design of coastal structures, waves are typically characterized by the significant wave height (Hs). Other essential wave parameters include the mean wave direction (θ) and the mean wave period (Tm), which are used to determine the response of coastal structures to incident wave conditions. Yuksel *et al.* (2020) noted that in addition to an increase in significant wave heights, a rise in the frequency of extreme events is likely in the future, which should be considered when determining design wave conditions. This increased likelihood of exceeding wave conditions can lead to larger wave forces, potentially causing damage to coastal structures.

Past climate can be analyzed using observational data (which is often limited in terms of time and spatial coverage) or wave modeling. To project future waves, information on future wind conditions is necessary. One approach involves using winds derived from general circulation models (GCMs) or regional climate models (RCMs) to drive a wave model (Chowdhury *et al.*, 2019). This enables the estimation of changes in the wave climate due to climate change.

Gumuscu *et al.* (2024) stated that extreme future wind climate for near and middle future periods under RCP4.5 scenario was significantly higher than the historical period in large return periods in the southwest of the island of Crete in the Eastern Mediterranean Sea. As a result, significant extreme wave heights for the 100-year return period were estimated to tend to increase by 5.85% over the historical period for both future periods in the Eastern Mediterranean Sea.

(iii) Design flow chart on performance-based design

The performance-based design philosophy and its concept are represented with a flow chart given in Fig. 2. In this context, the classification of port structures, the definition of damaged-based performance levels,

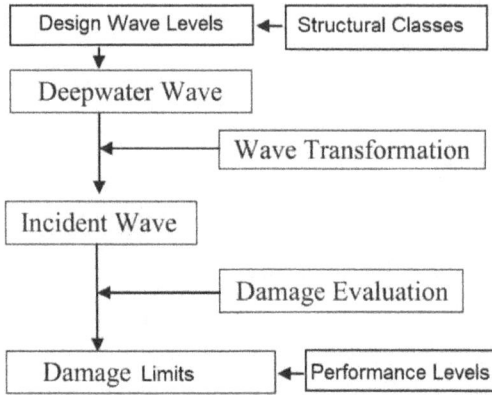

Fig. 2.   Flowchart for the performance-based design philosophy (Yuksel *et al.*, 2020).

multi-level design wave actions, and performance objectives are briefly described in line with the essential parameters of the performance-based design. As one of the essential parameters, the damage limit should be defined along with design wave parameters for the performance levels of each coastal structure.

Coastal structures are designed based on return levels derived from extreme value theory that provides a statistical description of the maxima of a stationary process, which assumes no change in the frequency of extremes over time. However, global climate change is expected to cause long-term changes in mean sea level, wave height, and storm frequency at time scales longer than the lifetime of many coastal structures. This causes a greater risk of damage to coastal structures than expected. Therefore, design wave parameters are required to be determined by considering the long-term historical data, the projected wave parameters, and the projected sea-level rise.

3. Coastal Resilience Measures

Burcharth *et al.* (2014a, 2014b) emphasized that climate change could result in rising sea levels and more intense storms. Both of these factors would increase the risk of flooding in low-lying areas, accelerate beach erosion, and damage existing coastal protection structures. To mitigate these challenges, it is crucial to upgrade these structures to meet or exceed their original design performance criteria. Upgrading might involve modifications to the structure's profile or the addition of new elements. Innovative design solutions will be essential to enhance coastal structures and ensure their resilience to the adverse effects of global climate change.

3.1. *Upgrading of conventional coastal breakwaters*

The stability number of rubble mound breakwaters is influenced by numerous parameters, including design wave height, unit shape, installation method, slope angle, and unit density. To withstand the effects of global climate change, existing or planned rubble mound breakwaters must be designed with resilience in mind.

A robust yet more environmentally friendly design is essential for rubble mound breakwaters and revetments due to several factors such as the following:

- Environmental impacts associated with quarrying operations,
- economic effects of construction and maintenance,
- larger protection layer units required due to climate change,
- aesthetic considerations of the protection layer,
- minimization of manufacturing and construction site costs,
- protection and improvement of ecological conditions.

By addressing these factors, innovative designs can be developed to enhance the sustainability and performance of rubble mound breakwaters and revetments.

As mentioned before, global climate change is characterized by rising sea levels and increasing intensity and duration of storms. These impacts can compromise the stability of breakwaters and other coastal protection structures, leading to heightened wave flooding. Mitigating these effects necessitates improvements to coastal defense structures, such as increasing structural stability and redesigning the crest level. However, these design requirements can result in increased costs and potentially environmentally unfriendly solutions.

Burcharth *et al.* (2014b) summarized their upgrading solutions, both with and without increasing the crest height of existing breakwaters, as depicted in Figs. 3 and 4. They also emphasized the importance of maintaining a good connection between the added armor layers and the existing structure, ideally using the same type and size of armor units. However, achieving good interlocking can be challenging, especially with

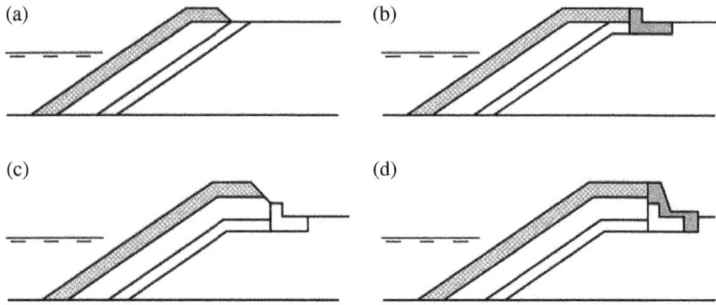

Fig. 3.   Increasing crest height case (Burcharth *et al.*, 2014b).

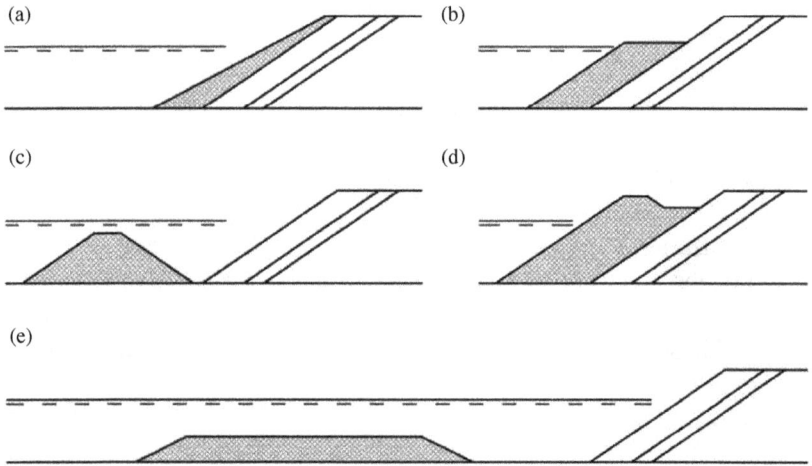

Fig. 4.   Improvement concept without increasing the crest (Burcharth *et al.*, 2014b).

complex armor unit types which require precise placement. In some cases, complete replacement of the armor units might be necessary.

Increasing the stability of breakwaters may require more complex solutions. While options such as adding larger units to the armor layer or increasing the slope angle might seem promising, Yuksel *et al.* (2022) demonstrated that these approaches may not significantly enhance stability. Alternative solutions should be thoroughly tested and existing expressions might be revisited. Moreover, optimization is also essential to compare costs and select the most efficient option.

Potential solutions for enhancing rubble mound breakwaters without increasing crest height might be as follows:

- Incorporating a berm slope. Design the structure with a sloped berm to improve stability and wave energy dissipation.
- Using high-density concrete blocks. Place high-density concrete blocks on the slope to enhance durability and reduce wave impact.
- Constructing a low-crested breakwater (tandem breakwaters). Add a low-crested breakwater in front of the existing structure to provide additional protection without altering the crest height.

One approach to strengthening existing breakwaters without increasing the crest height is to design a slope with a berm, creating a more resistant geometry against wave effects and reducing wave overtopping (Fig. 4(b) and 4(d)). Berm breakwaters, introduced in the early 1980s, are mass-armored structures that reshape into a more favorable S-shape due to wave action. The Icelandic type, a multi-layered structure with less reshaping, has gained popularity in areas with severe wave loading and abundant large stones.

Berm breakwaters can be categorized based on their reshaping behavior and construction method. PIANC (2003) proposed a classification based on reshaping behavior, while Van der Meer and Sigurdarson (2016) suggested a classification partly based on structural behavior, such as hardly reshaping, partly reshaping, and fully reshaping.

Numerous studies have focused on berm-type breakwaters in recent decades. A seaward slope berm is a cost-effective solution for reducing wave overtopping compared to increasing the crest height, and it can also enhance the stability of the armor layer (Burcharth and Frigaard, 1988; Van der Meer, 1988a; Hall and Kao, 1991; Lamberti *et al.*, 1994; Tomasicchio *et al.*, 1994; Van Gent, 1995; Lamberti and Tomasicchio, 1997; PIANC, 2003; Tørum *et al.*, 2003; Lykke Andersen, 2006; Lykke Andersen and Burcharth, 2010; Moghim and Tørum, 2012; Sigurdarson and Van der Meer, 2012, 2013; Van Gent, 2013; Sigurdarson *et al.*, 2013; Lykke Andersen *et al.*, 2014; Moghim and Alizadeh, 2014; Tørum and Sigurdarson, 2014; Van der Meer and Sigurdarson, 2014; Ehsani *et al.*, 2020; Yuksel *et al.*, 2020). Berm-type breakwaters can be particularly effective in adapting to increasing wave conditions due to climate change (Van Gent, 2019).

Berm-type breakwaters are classified based on their structural and hydraulic behavior, as defined by PIANC (2003), Van Gent (2013), and Van der Meer and Sigurdarson (2016). Within the concept of a dynamically stable rubble mound breakwater (Van der Meer, 1988a), the use of relatively small material in combination with a degree of seaward slope reshaping can result in reshaping berm-type breakwaters. These breakwaters can be classified as rubble mound breakwaters with a berm if they allow only a conventional amount of damage, as measured by Hudson's stability formula and Broderick's (1983) damage parameter S, to the armor layer.

Rubble mound breakwaters with a berm are generally more effective in enhancing the stability of the breakwater armor layer compared to conventional breakwaters. However, the potential damage at the transition between the lower slope and the berm is a critical factor in the stability of these structures. For rubble mound structures with a berm, the most vulnerable section is often the transition from the lower slope to the berm (Yuksel et al., 2018). While the berm width and berm level can significantly influence the stability of both the lower and upper slopes.

Van Gent (2013) also analyzed rubble mound structures consisting of a rock-armored slope and provided design formulas for non-reshaping or hardly reshaping mass armor-type breakwaters with a berm. He recommended further research to investigate the influence of a berm on the stability of concrete armor units in the lower slope and berm (see also the works of Van Gent and Van der Werf, 2017; Yuksel et al., 2020).

An alternative to conventional rubble mound armor units is the use of heavy concrete blocks. This approach can reduce the size of the concrete units and the required crane capacity. Yuksel et al. (2022) experimentally investigated the stability of high-density cube concrete blocks. Their study examined the performance of these high-density cubes in the armor layer of a breakwater. Two different cross-sections were modeled in a wave flume: one with a conventional cross-section and the other with a berm. Cubes with two different densities (24 kN/m³ for normal density (ND) and 31.5 kN/m³ for high density (HD)) and various placement methods were tested. The dimensions of the cubes were kept constant to avoid potential scale effects caused by differences in size. The results demonstrated that HD cubes were more stable than ND cubes. The characteristic wave height for HD cubes was 1.5 times higher than that of ND cubes. Using HD cubes can offer economic benefits, including reduced concrete volume, material transportation, and construction area requirements.

While a single layer of HD cubes proved to be highly stable, using HD cubes only in part of the armor layer or exclusively in the second (upper) layer resulted in decreased stability, especially compared to armor layers composed entirely of HD cubes.

Another approach to mitigating wave energy is to design a low-crest tandem breakwater on the offshore side of the conventional breakwater, as illustrated in Fig. 4(c) and 4(e). This configuration can effectively reduce the wave energy reaching the conventional breakwater, providing it with additional protection against high-energy waves (Thesnaar, 2015).

The increasing water levels, wave heights, and storm frequencies associated with global climate change highlight the need for more resilient coastal structures, adaptation measures, and improvements in low-crested breakwater design. These structures can effectively protect coastal areas from erosion and offer advantages over conventional non-overtopped structures, such as being more environmentally friendly and economical. As a result, their use is expected to increase in the future.

When structures are sufficiently high to prevent overtopping, the armor on the crest and rear slope can be smaller than that on the front slope. Such structures are typically referred to as conventional rubble mound breakwaters. However, many structures are designed to experience some or even severe wave overtopping under design conditions. Conversely, some structures are so low that they are overtopped even during daily wave conditions. Structures with crest levels near or below still water level will inevitably be subjected to overtopping and wave transmission.

When a structure's crest level is low, wave energy can pass over the structure. This has two primary effects:

1. **Reduced armor requirements on the front slope:** Less energy is transmitted to the front slope, resulting in lower run-down forces, which allows for the use of smaller armor units.
2. **Necessity for armoring the crest and rear slope:** These areas must be armored to withstand the impact of overtopping waves (Yuksel *et al.*, 2025).

Low-crested rubble mound breakwaters (LCBs) can be categorized into three types: dynamically stable reef breakwaters, statically stable low-crested emerged type structures, and statically stable submerged structures.

Low-crested breakwaters have been extensively studied for many years, with numerous studies focusing on their stability, wave transmission, and structural response. Research on the structural response of low-crested breakwaters has often been conducted under specified boundary conditions. These studies primarily concentrated on structures with rock armor layers (Givler and Sørensen, 1986; Ahrens, 1987, 1989; Van der Meer, 1990; Vidal *et al.*, 1992, 1995; Burger, 1995; Burcharth *et al.*, 2006; Muttray *et al.*, 2012; Yuksel *et al.*, 2025).

The hydraulic response of low-crested breakwaters involves a complex interplay of wave transmission, reflection, and energy dissipation. Optimizing these structures requires careful consideration of design parameters, environmental impacts, and the dynamic marine environment. In this type of breakwater, wave transmission occurs through wave overtopping and wave propagation from the permeable structure to the rear side. The amount of wave overtopping is primarily influenced by the crest freeboard and crest width.

Low-crested breakwaters (LCBs) reflect a portion of incident wave energy seaward. The reflection coefficient is influenced by factors, such as the structure's geometry, surface roughness, and incident wave conditions. LCBs generally reflect less energy than high (non-overtopped) crested breakwaters. The permeable nature and armor layer selection of LCBs, often involving rubble mound structures, contribute to energy dissipation through friction and turbulence, further reducing energy transmission to the rear slope. Extensive research has been conducted on the hydraulic behavior of low-crested breakwaters, with ongoing studies contributing to our understanding of these structures (Van der Meer and Pilarczyk, 1990; Daemen, 1991; Ahrens, 1987; Van der Meer and Daemen, 1994; d'Angremond *et al.*, 1996; Seabrook and Hall, 1998; Calabrese *et al.*, 2002; Briganti *et al.*, 2003; Van der Meer *et al.*, 2005; Buccino and Calabrese, 2007; Goda and Ahrens, 2008; Tomasicchio and D'Alessandro, 2013; Zanuttigh and Van der Meer, 2008; Zhang and Li, 2014; Sindhu and Shirlal, 2015; Giantsi and Moutzouris, 2016; Kurdistani *et al.*, 2022; Van Gent *et al.*, 2023; Yuksel *et al.*, 2025).

3.2. *Nature-based solutions*

Building with nature is a comprehensive engineering approach that leverages natural ecological processes to achieve sustainable coastal structure designs.

As emphasized by Airoldi *et al.* (2005), coastal areas play a crucial role in the economic, social, and political development of many nations. These areas support diverse and productive ecosystems that provide valuable goods and services. However, coastal flooding and erosion pose significant global threats, exacerbated by human-induced changes and accelerated sea-level rise. Over the past century, hard coastal defense structures have become a common response to these threats, significantly altering coastal landscapes. In some regions, these structures now dominate the shoreline, leading to substantial environmental changes. Despite this, the ecological impacts of coastal defense have often been overlooked. These impacts are evident locally, where defense structures disrupt soft-bottom environments and introduce artificial hard-bottom habitats, altering native species assemblages. Regionally, the proliferation of coastal defense structures can critically impact species diversity by removing natural structures, promoting the spread of non-native species, and increasing habitat heterogeneity.

Understanding the environmental context of these structures is important for effective management, as their impacts can be both general and site-specific. To achieve management goals, specific strategies are needed to mitigate environmental impacts, such as minimizing changes to surrounding sediments, controlling the spread of exotic or nuisance species, and enhancing natural resources such as fish recruitment or promoting biodiversity for eco-tourism.

The impact of climate change on the functionality of nature-based coastal protection remains uncertain. To safely apply this protection measure, a thorough understanding of the wave attenuation process in coastal wetlands and related predictive models is necessary. Nature-based coastal protection that integrates vegetated wetlands for wave attenuation and erosion mitigation shows great promise (Huang *et al.*, 2024).

To enhance the ecological value of vertical hard coastal structures, hybrid designs with complex surface textures (such as a combination of grooves and pits) have been recommended (Fig. 5). This strategy optimizes ecological colonization at two spatial scales: 1. At the millimeter scale, barnacle abundance, known for its bioprotective capabilities, is promoted, 2. At the centimeter scale, species richness and abundance are increased through the incorporation or creation of habitat features (MacArthur *et al.*, 2019). By incorporating these complex surface textures, hybrid designs can foster greater biodiversity and ecological resilience on coastal structures.

Fig. 5. Living seawalls with marine habitats in Sydney Harbor. [https://www. climatechange.environment.nsw.gov.au/stories-and-case-studies/living-seawalls].

3.2.1. *Criteria for coastal nature-based solutions*

To ensure the ecological integrity of coastal areas while implementing nature-based design solutions, the following criteria can be defined:

1. Impact classification

Coastal and marine areas can be classified based on the degree of human influence as follows:

- **High Human Impact:** Areas significantly altered by human activities, such as industrialization, urbanization, or agriculture. These areas often experience habitat degradation, invasive species introduction, and biodiversity loss.
- **Moderate Human Impact:** Areas influenced by both human activities and natural processes. These areas may face moderate pollution, habitat alteration, and limited resilience to natural disturbances.
- **Low Human Impact:** Areas primarily shaped by natural processes with minimal human intervention. While these areas may still be affected by factors such as climate change or natural disasters, they generally retain a high degree of ecological integrity.

By classifying coastal areas based on their impact level, decision-makers can tailor nature-based design solutions to address specific challenges and promote sustainable coastal management.

2. Problem identification and site characterization

Effective coastal and marine management requires a comprehensive understanding of specific challenges and site characteristics. Key areas of concern include the following:

- **Morphological Changes:** Erosion, sedimentation, and sea-level rise, both natural and human-induced, can significantly alter coastal landscapes.
- **Ecological Deterioration:** Habitat destruction, pollution, invasive species, and climate change impacts threaten the health of coastal and marine ecosystems.
- **Water Quality Issues:** Pollution from industrial discharge, agricultural runoff, sewage, and litter can degrade water quality and harm marine life.
- **Infrastructure Vulnerability:** Coastal infrastructure, including ports, harbors, and coastal defenses, faces risks from storm surges, sea-level rise, and erosion.
- **Resource Depletion:** Overfishing, habitat destruction, and pollution contribute to the decline of marine resources like fish stocks and coral reefs.
- **Tourism Impacts:** Coastal and marine tourism can be affected by environmental degradation, water pollution, and development pressures.
- **Biodiversity Loss:** Human activities threaten species diversity and ecosystem resilience in coastal and marine regions.

By carefully identifying and characterizing these issues, decision-makers can develop targeted management strategies to address specific challenges and promote the long-term sustainability of coastal and marine environments.

3. Problem parameterization and site-based modeling

Understanding and predicting the impacts of environmental factors on coastal regions requires a robust framework of problem parameterization and site-based modeling. Key parameters to consider include the following:

- **Sea-Level Rise:** Rising sea levels pose threats to coastal communities, including inundation, erosion, and saltwater intrusion.
- **Water Temperature Increase:** Warmer ocean temperatures can lead to coral bleaching, species shifts, and changes in ocean circulation.

- **Atmospheric Pressure Field Changes:** Shifts in atmospheric pressure influence weather patterns, ocean circulation, and storm behavior.
- **Storm Intensification and Frequency:** Climate change is increasing the intensity and frequency of storms, posing risks to coastal areas.
- **Wave Climate Changes:** Alterations in wave height, direction, and frequency impact coastal erosion, sediment transport, and habitats.

As climate change intensifies, extreme wave events are becoming more frequent and severe, posing significant risks to coastal communities, ecosystems, and infrastructure. To mitigate these risks, accurate forecasting of extreme waves is crucial.

Advanced Forecasting Methods:

i. **Climate Models:** Incorporating rising sea levels, increasing ocean temperatures, and changing wind patterns, advanced climate models can provide valuable insights into future wave conditions.
ii. **Regional Variations:** Recognizing that coastal areas experience distinct impacts based on local geography and ocean currents, these models must account for regional variations.
iii. **High-Resolution Data:** Integrating high-resolution satellite data and real-time ocean observations enhances the accuracy of wave forecasts.

Applications of Wave Forecasting:

i. **Coastal Defense:** Developing effective coastal defense strategies, such as breakwaters, seawalls, and beach nourishment projects.
ii. **Early Warning Systems:** Improving early warning systems to alert communities of impending extreme wave events.
iii. **Infrastructure Planning:** Ensuring the resilience of coastal infrastructure, including ports, harbors, and offshore energy facilities.

Addressing Wave Climate Changes:

i. **Understanding Impacts:** Characterizing wave climate variability to assess coastal erosion rates and identify vulnerable areas.
ii. **Designing Protection Measures:** Developing tailored coastal protection measures to mitigate the effects of extreme waves.

By leveraging advanced forecasting techniques and understanding the impacts of extreme waves, coastal communities can better prepare for and adapt to the challenges posed by climate change.

Site-based modeling enables the following:

- **Quantifying the impacts of these factors:** Assessing the extent of inundation, thermal stress on marine organisms, storm surge risks, and coastal erosion rates.
- **Informing decision-making:** Developing adaptation strategies, such as coastal defenses, managed retreats, marine conservation measures, and disaster risk reduction plans.
- **Promoting sustainable coastal management:** Ensuring the resilience of coastal communities and ecosystems in the face of climate change.

By integrating problem parameterization and site-based modeling, decision-makers can better understand and address the complex challenges facing coastal regions.

4. Evaluating possible solutions

Evaluating potential solutions for coastal and marine challenges requires a comprehensive approach that considers the following:

- **Problem-Specific Factors:** Identifying the root causes of each problem to tailor solutions effectively.
- **Multifaceted Approach:** Combining policy interventions, regulatory measures, community engagement, and stakeholder collaboration.
- **Sustainability:** Prioritizing solutions that balance ecological, social, and economic considerations.

Sustainable Planning and Design

To achieve sustainable outcomes, solutions must be as follows:

- **Site-Specific:** Tailored to the unique characteristics of each coastal area.
- **Resilient:** Capable of adapting to changing conditions and future challenges.
- **Nature-Based:** Integrating natural processes and ecosystems to enhance sustainability.

By adopting a comprehensive and sustainable approach, decision-makers can develop solutions that promote the health, resilience, and well-being of coastal and marine environments.

5. Defining design parameters

Effective design of coastal and marine solutions requires a comprehensive understanding of complex environmental interactions. This involves the following:

- **Multidisciplinary Approach:** Combining scientific research, modeling, and field studies.
- **Environmental Modeling:** Using advanced tools to simulate coastal processes and evaluate potential impacts.
- **Field Studies and Monitoring:** Collecting data on coastal dynamics, ecological responses, and human interactions.
- **Stakeholder Engagement:** Involving local communities, government agencies, NGOs, and industry stakeholders in the design process.
- **Risk Assessment:** Identifying potential hazards and vulnerabilities associated with proposed solutions.
- **Cost-Benefit Analysis:** Evaluating the economic feasibility and societal value of alternative solutions.

By integrating these elements, decision-makers can define design parameters that are as follows:

- **Evidence-Based:** Informed by scientific research and data.
- **Adaptive:** Capable of adjusting to changing conditions.
- **Sustainable:** Balancing ecological, social, and economic considerations.

This comprehensive approach ensures that coastal and marine interventions are tailored to specific needs, address challenges effectively, and maximize benefits while minimizing negative impacts.

6. Selection of nature-based solutions for coastal and marine challenges

To address the complex challenges facing coastal areas and oceans, nature-based solutions are emerging as a promising approach. These solutions leverage natural processes and ecosystems to achieve sustainable outcomes while maintaining the integrity of coastal morphology and ecosystems.

Key objectives of nature-based solutions:

- **Increase Coastal Resilience:** Increase Coastal Resilience enhances the ability of coastal areas to withstand erosion, storm surges, and sea-level rise.
- **Improve Water Quality:** Improve Water Quality reduces pollution and improve the quality of coastal waters.
- **Support Biodiversity:** Support Biodiversity protects and restores biodiversity in coastal and marine ecosystems.
- **Promote Carbon Sequestration:** Promote Carbon Sequestration contributes to carbon storage and climate mitigation.

Types of Nature-Based Solutions:

- **Hard Solutions:** Engineering interventions that mimic or strengthen natural coastal features, such as breakwaters, seawalls, or living shorelines.
- **Soft Solutions:** Ecosystem-based approaches that utilize natural processes, such as wetlands, oyster reefs, or mangrove forests, to protect coastlines.
- **Hybrid Solutions:** Combining hard and soft measures for a more comprehensive approach.

3.2.2. *Nature-based hard measures*

Nature-based hard measures are engineering interventions designed to protect shorelines while supporting ecosystem health. They often incorporate natural materials and mimic natural coastal features, creating durable and sustainable defenses. Examples include the following:

- **Living Shorelines:** Using vegetation, rocks, or oyster reefs to stabilize shorelines and provide habitat for marine life.
- **Biodegradable Breakwaters:** Constructing breakwaters from biodegradable materials that decompose over time, reducing their environmental impact.

- **Hybrid Structures:** Combining traditional engineering techniques with natural elements, such as using oyster reefs as foundations for seawalls.

Nature-based hard structures offer innovative solutions for coastal protection that integrate engineering expertise with ecological principles. These structures provide resilience, biodiversity benefits, and improved water quality while maintaining a more natural aesthetic.

Specific examples of nature-based hard measures:

  i. **Vegetated Seawalls:** Combining seawalls with vegetation can stabilize sediments, improve ecological connectivity, and provide wildlife habitat.
 ii. **Gabion Baskets and Rock Linings:** These structures, when combined with vegetation, offer erosion resistance and create natural-looking barriers.
iii. **Breakwaters with Artificial Reefs:** Integrating artificial reefs into breakwaters can provide additional habitat for marine life and enhance biodiversity.
 iv. **Groins with Beach Nourishment:** Groins and beach nourishment can help restore and protect beaches while supporting coastal ecosystems.
  v. **Biogenic Reefs:** Living organisms such as oysters and mussels can form reefs that reduce wave energy, filter water, and provide habitat.
 vi. **Submerged Breakwaters:** These structures can be designed to include ecological features that benefit marine life.

Advantages of nature-based hard structures:

  i. **Increased Resilience:** Better adaptation to changing environmental conditions.
 ii. **Enhanced Biodiversity:** Creation of new habitats for marine organisms.
iii. **Improved Water Quality:** Filtration of pollutants and improved water clarity.
 iv. **Aesthetic Appeal:** More natural appearance compared to traditional structures.

Considerations for Implementation:

i. **Site-Specific Needs:** Each coastal area has unique characteristics that must be considered when selecting and designing nature-based hard structures.
ii. **Potential Challenges:** Careful evaluation of potential risks and challenges is essential for successful implementation.

By carefully considering these factors, nature-based hard structures can offer a promising approach to coastal protection that provides both environmental and economic benefits.

3.2.3. *Nature-based soft measures*

Nature-based soft coastal measures are ecological and habitat-focused strategies that provide resilience, erosion protection, and storm surge defense. Unlike hard structures, they leverage natural processes and ecosystems to achieve these goals:

- **Beach Dune Restoration:** Beach dune restoration involves the rehabilitation of natural dunes along the coastline to protect against erosion and storm damage. This approach often involves planting native dune vegetation, such as beach grasses and shrubs, to stabilize the sand and create natural barriers against wind and wave action. Restored dune systems can also provide habitat for wildlife and contribute to the resilience of coasts by absorbing wave energy and reducing the effects of storm surges.

Dune nourishment also involves filling eroded beaches and dunes with sand to restore coastal habitats and protect them from erosion. This gentle measure mimics natural sediment transport processes and promotes beach deposition, increasing coastal resilience and providing habitat for beach-dwelling species. Dune nourishment projects may involve dredging offshore sand deposits or importing sand from inland sources to rebuild eroded shorelines:

- **Salt Marsh Creation and Restoration:** Salt marshes are coastal wetlands characterized by unique vegetation and tidal influence.

Creating and restoring salt marsh habitats can help stabilize shore-lines, trap sediments, and provide critical habitat for fish and wildlife. Salt marshes also offer natural flood mitigation benefits by absorbing and dissipating wave energy during storms. Restoration efforts may include restoring tidal flow, removing invasive species, and planting native vegetation to increase ecosystem health and resilience.

- **Living Shorelines:** Living shorelines are nature-based approaches to coastal stabilization that include natural features such as marsh veg-etation, oyster reefs, and submerged aquatic vegetation. These gentle measures can use natural vegetation, such as marsh grasses and man-groves, to stabilize sediments, reduce wave energy, and provide habi-tat for fish and wildlife. Living shorelines can help reduce erosion, improve water quality, and increase biodiversity along coastlines. They also offer social and recreational benefits, such as nature-based tourism and outdoor recreation opportunities.

- Seagrasses are underwater flowering plants that form extensive mead-ows in shallow coastal waters. They play a crucial role in stabilizing sediments, reducing wave energy, and mitigating erosion. Seagrass meadows also provide essential food and shelter for various marine life, contributing to healthy coastal ecosystems (Fig. 6). Seagrass meadows, often referred to as underwater grasslands, are vital ecosys-tems found in shallow marine waters. They play a crucial role in carbon storage, which has significant implications for climate change mitigation. However, Lei *et al.* (2023) indicated that the majority of the carbon stored in seagrass sediments originates outside the meadow, such that the carbon storage capacity within a meadow is strongly dependent on hydrodynamic conditions that favor deposition

Fig. 6.   Ocean floor with seagrass [https://oceanfdn.org/seagrass/].

and retention of fine organic matter within the meadow. By extension, if hydrodynamic conditions vary across a meadow, they may give rise to spatial gradients in carbon.

- **Mangrove Afforestation:** Mangrove forests are highly productive coastal ecosystems that provide valuable ecosystem services, including shoreline stabilization, carbon sequestration, and habitat provision. Reforestation efforts aim to restore degraded mangrove habitats by replanting mangrove seedlings and implementing measures to protect existing mangrove stands. Mangrove forests help reduce coastal erosion, trap sediments, and provide a breeding habitat for fish and other marine species.

- Oyster reefs are biogenic structures formed by the accumulation of living and dead oyster shells. They excel at attenuating wave energy, protecting shorelines from erosion, and improving water quality through filtration. Oyster reefs also provide critical habitat for diverse marine organisms, promoting biodiversity and ecosystem health.

Benefits of nature-based soft coastal measures:

   i. **Increased Resilience:** Enhanced ability to withstand erosion, storms, and sea-level rise.
  ii. **Improved Ecosystem Health:** Protection of biodiversity and habitat for marine life.
 iii. **Water Quality Benefits:** Filtration of pollutants and improved water clarity.
  iv. **Carbon Sequestration:** Contribution to climate change mitigation.
   v. **Social and Recreational Benefits:** Support for tourism, recreation, and coastal communities.

By harnessing the power of natural ecosystems, nature-based soft coastal measures offer sustainable and effective solutions to protect coastlines and promote environmental health.

3.2.4. *Nature-based hybrid measures*

Conventional hard and soft measures for coastal areas can be integrated with ecologically based solutions. However, integrated measures should consider the ecological life in the problematic regions.

Integrating conventional hard and soft measures with ecologically based solutions can provide a more comprehensive and effective approach to coastal protection. However, it's crucial to carefully consider the specific ecological conditions and life within the problematic regions. Some key factors may be considered when integrating hard and soft measures:

i. **Habitat Compatibility:** Habitat Compatibility ensures that the chosen measures do not disrupt or damage existing habitats.
ii. **Ecological Connectivity:** Ecological Connectivity promotes connectivity between different habitats to support biodiversity.
iii. **Long-Term Sustainability:** Long-Term Sustainability evaluates the long-term ecological impacts of the integrated measures.
iv. **Adaptive Management:** Adaptive Management implements a monitoring and evaluation framework to adapt measures as needed.

By carefully considering these factors, it's possible to develop integrated solutions that provide both protection and environmental benefits for coastal areas.

Natural features are formed over time through the interplay of physical, geological, biological, and chemical processes. As nature-based solutions aim to address specific ecological or ecosystem functions, they are often tailored to particular sites and scenarios. Their design requires expertise across various fields, including coastal ecology, geology, oceanography, and engineering.

The Federal Highway Administration (FHWA, 2019) highlights examples of nature-based solutions, such as tidal marshes, mangroves, maritime forests, reefs, beaches, and dunes. According to FHWA (2019), these solutions can mitigate threats to coastal infrastructure by reducing storm surge flooding, wave damage, erosion, shoreline retreat, and potential sea-level rise impacts. A nature-based solution may consist entirely of natural elements (e.g., vegetation, beach, and dune) or incorporate a combination of natural, constructed natural elements, and traditional coastal structures (e.g., sill, breakwater, revetment, and seawall). A hybrid approach combines natural and constructed elements. FHWA (2019) notes that as conditions vary — such as the steepness of slopes, wave climate, and the degree of shoreline exposure — nature based solutions increasingly rely on coastal structures or armoring. For instance, as exposure to large waves or the need for risk reduction increases, more substantial structures may be necessary, such as a sill near the shore, a breakwater

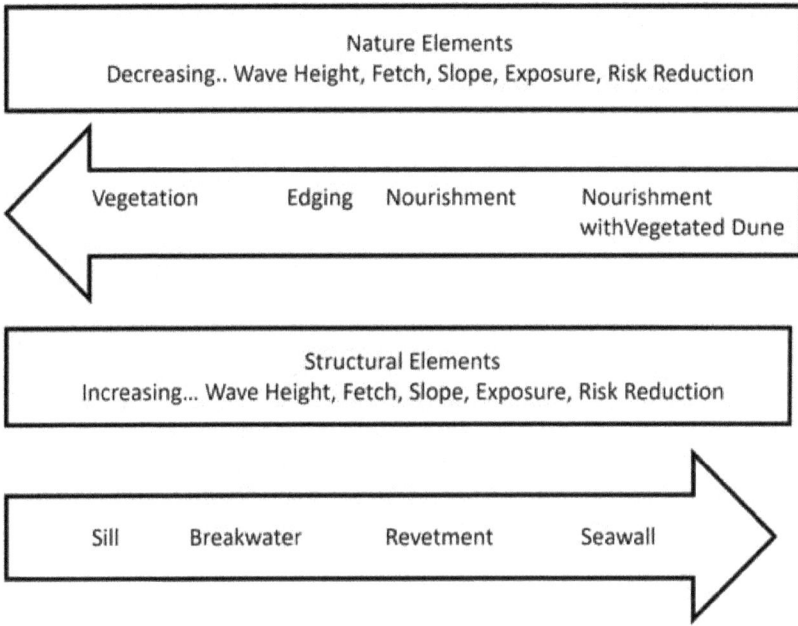

Fig. 7. Nature-based solutions consist of varying degrees of natural, nature-based, and structural elements depending on the setting, exposure to wave action, and resilience needs (revised from FHWA, 2019).

farther out, or a continuous rock revetment or seawall. Conversely, when wave exposure or the required level of risk reduction decreases, a suitable nature-based solution might focus more on natural features or engineered nature-based features with minimal structural components. The examples provided in Fig. 7 include beach nourishment with and without vegetated dunes, edging (a form of low shoreline bank stabilization typically used with newly placed sediment and vegetation), and shoreline or upland vegetation stands.

The FHWA (2019) concluded that the choice between a purely nature-based approach, a fully structural approach, or a hybrid solution depends on several factors, including the following:

- Resilience needs or the level of risk reduction required,
- ecological and geological context,
- exposure to large waves,

- project objectives,
- project cost,
- desired level of reliability,
- local policies and regulations.

References

Ahrens, J. P. (1987). Characteristics of reef breakwaters, *Technical Report CERC* pp. 87–17, US Army Corps of Engineers.

Ahrens, J. P. (1989). Stability of reef breakwaters. *Journal of Waterway, Port, Coastal, and Ocean Engineering, ASCE,* 115(2), 221–234.

Airoldi L., Abbiati M., Beck M. W., Hawkins S. J., Jonsson P. R., Martin D., Moschella, P. S., Sundelo A., Thompson R. C., Aberg, P. (2005). An ecological perspective on the deployment and design of low-crested and other hard coastal defence structures. *Coastal Engineering,* 52, 1073–1087. DOI: 10.1016/j.coastaleng.2005.09.007.

Alvarez, F. and Lorenzo, J. (2019). Regional climate models and their application in coastal areas. *Environmental Research Letters,* 14(8), 084020. DOI: 10.1088/1748-9326/ab26b1.

Ashton, A. D., Wright, S., and Moser, S. (2008). Coastal impacts of sea-level rise: A review of modeling approaches and challenges. *Journal of Coastal Research,* 24(2), 371–380. DOI: 10.2112/06-0807.1.

Bergillos, R. J., Rodríguez-Gómez, S., and Martínez-del-Pozo, J. A. (2019). Coastal erosion and flooding hazard assessment: A case study in the Gulf of Cádiz. *Journal of Coastal Research,* 35(6), 1174–1187. DOI: 10.2112/JCOASTRES-D-19-00068.

Briganti, R., Van der Meer, J., Buccino, M., and Calabrese, M. (2003). Wave transmission behind low-crested structures. *International Conference on Coastal Structures 2003,* Portland, Oregon, USA, (pp. 580–592). https://doi.org/10.1061/40733(147)48.

Broderick, L. L. (1983), Rip rap stability, a progress report. In *Proceedings of the Coastal Structures,* 83, 320–330. ASCE.

Brown, W., Munk, W., Snodgrass, F., Mofjeld, H., and Zetler, B. (1975). Mode bottom experiment. *Journal of Physical Oceanography,* 5, 75–85. https://doi.org/10.1175/1520-0485(1975)005<0075:MBE>2.0.CO;2

Buccino, M. and Calabrese, M. (2007). Conceptual approach for prediction of wave transmission at low-crested breakwaters. *Journal of Waterway, Port, Coastal, and Ocean Engineering,* ASCE, 133(3), 213–224. https://doi.org/10.1061/(ASCE)0733-950X(2007)133:3(213).

Burcharth, H. F. and Frigaard, P. (1988). On 3-dimensional stability of reshaping breakwaters. *Proceedings 21st International Conference on Coastal Engineering* (ICCE1988).

Burcharth, H. F., Kramer, M., Lamberti, A., and Zanuttigh, B. (2006). Structural stability of detached low crested breakwaters. *Coastal Engineering*, 53(4), 381–394. https://doi.org/10.1016/j.coastaleng.2005.10.023.

Burcharth, H. F., Lamberti, A., and Røsnes, R. (2014a). Coastal structures and climate change: Adaptation and resilience. *Coastal Engineering*, 87, 122–132. DOI: 10.1016/j.coastaleng.2014.01.005.

Burcharth, H. F., Lykke Andersen, T., and Lara, J. L. (2014b). Upgrade of coastal defence structures against increased loadings caused by climate change: A first methodological approach. *Coastal Engineering*, 87, 112–121. http://dx.doi.org/10.1016/j.coastaleng.2013.12.006.

Burger, G. (1995). Stability of breakwaters with low crest (in Dutch), Master thesis, TU Delft.

Calabrese, M., Vicinanza, D., and Buccino, M. (2002). Large-scale experiments on the behaviour of low crested and submerged breakwaters in presence of broken waves. *28th International Conference on Coastal Engineering (ICCE) 2002*, Cardiff, Wales, pp. 1900–1912, World Scientific. DOI: 10.1142/9789812791306_0160.

Chowdhury, M. R., Islam, M. N., and Ali, M. M. (2019). Oceanic response to changing climate: Implications for coastal management. *Journal of Marine Systems*, 191, 1–12. DOI: 10.1016/j.jmarsys.2019.01.004.

Chowdhury, P., Behera, M. R., and Reeve, D. E. (2019). Wave Climate Projections along the Indian Coast. *International Journal of Climatology*, 39(11), 4531–4542.

Daemen, I. F. R. (1991). Wave Transmission at Low Crested Structures; Master thesis, TU Delft. http://resolver.tudelft.nl/uuid:433dfcf3-eb87-4dc9-88dc-8969996a6e3f.

d'Angremond, K., Van der Meer, J. W., and De Jong, R. J. (1996). Wave transmission at low-crested structures. *25th International Conference on Coastal Engineering (ICCE) 1996*, Orlando, Florida, pp. 2418–2427. https://doi.org/10.1061/9780784402429.187.

Do, T. Q., Van de Lindt, J. W., and Cox, D. T. (2016). Performance-based design methodology for inundated elevated coastal structures subjected to wave load. *Engineering Structures*, 117, 250–262.

Ehsani, M., Moghim, M. N., and Shafieefar, M. (2020). An experimental study on the hydraulic stability of icelandic-type berm breakwaters. *Coastal Engineering*, 156, 1–17.

FHWA (The Federal Highway Administration). (2019). Nature-Based Solutions for Coastal Highway Resilience: An Implementation Guide. Washington, DC, USA.

Giantsi, T. and Moutzouris, C. I. (2016). Experimental investigation of wave transmission and reflection at a system of low crested breakwaters. *13th International Conference on Protection and Restoration of the Environment.* https://doi.org/10.1007/s40710-017-0237-8.

Givler, D. L. and Sorensen, R. M. (1986). An investigation of the stability of submerged homogeneous rubble-mound structures under wave attack. Report No. HL-110-86, H. R. Imbt Hydraulics Laboratory, Department of Civil Engineering, Lehigh University, Bethlehem, PA,

Goda, Y. (2004). Spread parameter of extreme wave height distribution for performance-based design of maritime structures. *Journal of Waterway, Port, Coastal, and Ocean Engineering*, 130(1), 29–38.

Goda, Y. and Ahrens, J. P. (2008). New formulation of wave transmission over and through low-crested structures. *31th International Conference on Coastal Engineering (ICCE) 2008*, Hamburg, Germany, 5, pp. 3530–3541. World Scientific. https:// doi.org/10.1142/7342.

Goda, Y. and Takahashi, S. (2001). Advanced design of maritime structures in the 21st century, Port and Harbour Research Institute, Yokosuka, Japan.

Gumuscu, I. (2024). Modeling the Effect of Global Climate Change on Wind and Wave Climate in the Eastern Mediterranean Basin, revised from 4th progress report of PhD. Thesis, YTU, (in Turkish), in progress.

Gumuscu, I., Sahin, C., Yuksel, Y., Ari Guner, H. A. and Islek, F. (2024). Evaluation of Future Wind Climate over the Eastern Mediterranean Sea. *Regional Studies in Marine Science*, 78, 103780. DOI: 10.1016/j.rsma.2024. 103780.

Hall, K. R. and Kao, J. S. (1991). The influence of armour stone gradation on dynamically stable breakwaters. *Coastal Engineering*, 15, 333–346.

Hdidouan, B. and Staffell, I. (2017). The impact of changing wind characteristics on wind energy production. *Renewable Energy*, 101, 43–56. DOI: 10.1016/ j.renene.2016.09.033.

Hemer, M. A., Mori, N., and Wang, X. (2013). Future changes in extreme wave conditions across the globe. *Geophysical Research Letters*, 40(24), 6261–6267. DOI: 10.1002/2013GL058764.

IPCC. (2021). Climate Change 2021: *The Physical Science Basis*. Cambridge University Press. DOI: 10.1017/9781009157896.

IPCC. (2023). Climate Change 2023: *The Physical Science Basis*. Cambridge University Press. DOI: 10.1017/9781009328492.

Islek, F. and Yuksel, Y. (2023). Spatio-temporal long-term evaluations of the mean sea level pressure, sea level change, and sea surface temperature over two enclosed seas. *Regional Studies in Marine Science*, 1–23. DOI: 10.1016/ j.rsma.2023.103130.

Islek, F., Yuksel, Y., and Sahin, C. (2022a). Evaluation of regional climate models and future wind characteristics in the Black Sea. *International Journal of Climatology*, 42, 1877–1901. DOI: 10.1002/joc.7341.

Islek, F., Yuksel, Y., and Sahin, C. (2022b). Evaluation of regional climate models and future wave characteristics in an enclosed sea: A case study of the Black Sea. *Ocean Engineering*, 262, 112220. DOI: 10.1016/j.oceaneng. 2022.112220.

Komar, P. D. (2011). Coastal erosion processes and impacts: The consequences of earth's changing climate and human modifications of the environment, *Treatise on Estuarine and Coastal Science.* https://doi.org/10.1016/B978-0-12-374711-2.00314-4.

Kurdistani, S. M., Tomasicchio, G. R., Felice, D., and Francone, A. (2022). Formula for wave transmission at submerged homogeneous porous breakwaters. *Ocean Engineering*, 266, 113053. https://doi.org/10.1016/j.oceaneng.2022.113053.

Lamberti, A. and Tomasicchio, G. R. (1997). Stone mobility and longshore transport at reshaping breakwaters. *Coastal Engineering* 29, 263–289.

Lamberti, A., Tomasicchio, G. R., and Guiducci, F. (1994). Reshaping breakwaters in deep and shallow water conditions. *Proceeding 24th International Conference on Coastal Engineering (ICCE 1994). ASCE*, 1343–1358.

Lemos, M. C., Lo, J., and Sinsky, E. (2019). Modeling climate data for future projections: A review. *Climate Dynamics*, 53(1–2), 371–391. DOI: 10.1007/s00382-019-04720-7.

Ling, H. I., Cheng, A. H. D., Mohri, Y., and Kawabata, T. (1999). Permanent displacement of composite breakwaters subject to wave impact. *Journal of Waterway, Port, Coastal, and Ocean Engineering*, 125(1), 1–8.

Lykke Andersen, T. (2006). Hydraulic Response of Rubble Mound Breakwaters. Scale Effects — Berm Breakwaters. PhD Thesis. Department of Civil Engineering, Aalborg University, Denmark.

Lykke Andersen, T. and Burcharth, H. F. (2010). A new formula for front slope recession of berm breakwaters. *Coastal Engineering,* 57, 359–374.

Lykke Andersen, T., Moghim, M. N., and Burcharth, H. F. (2014). Revised Recession of Reshaping Berm Breakwaters. *Proceeding of 34rd International Conference on Coastal Engineering*, pp. 15–20, June 2014, Seoul.

MacArthur, M., Naylor, L. A., Hansom, J. D., Burrows, M. T., Loke, L. H. L., Boyd I. (2019). Maximising the ecological value of hard coastal structures using textured formliners. *Ecological Engineering.* https://doi.org/10.1016/j.ecoena.2019.100002.

Meucci, A., Benassi, G., and Camargo, S. J. (2020). Climate models and their role in future projections: Insights and advancements. *Journal of Climate*, 33(10), 4101–4120. DOI: 10.1175/JCLI-D-19-0667.1.

Meinshausen, M., Smith, S. J., Calvin, K., Daniel, J. S., Kainuma, M. L. T., Lamarque, J.-F., Matsumoto, K., Montzka, S. A., Raper, S. C. B., Riahi, K., Thomson, A., Velders, G. J. M., and van Vuuren, D. P. (2011). The RCP greenhouse gas concentrations and their extensions from 1765 to 2300, *Climatic Change,* 109, 213–241. https://doi.org/10.1007/s10584-011-0156-z.

Moghim, M. N. and Alizadeh, F. (2014). Hydraulic stability of reshaping berm breakwaters using the wave momentum flux parameter. *Coastal Engineering*, 83, 56–64.

Moghim, M. N. and Tørum, A. (2012). Wave induced loading of the reshaping rubble mound breakwaters. *Applied Ocean Research*, 37, 90–97.

Muttray, M., Oumeraci, H., and Ten Oever, E. (2006). Wave reflection and wave run-up at rubble mound breakwaters. *30th International Conference on Coastal Engineering (ICCE) 2006*, San Diego, California, USA, Vol. 5, pp. 4314–4324. World Scientific. https://doi.org/10.1142/9789812709554_0362.

Overseas Coastal Area Development Institute of Japan (OCDI). 2009. Technical standards and commentaries for port and harbor facilities in Japan, OCDI, Tokyo.

PIANC, (2001). Seismic Design Guidelines for Port Structures.

PIANC, (2003). State of the Art of Designing and Constructing Berm Breakwaters, WG 40, Brussels.

PIANC, (2014). Countries in Transition (CIT): Coastal Erosion Mitigation Guidelines. Report no 123.

Ranasinghe, R., Callaghan, D. P., and Stive, M. J. F. (2012). Variability and trends in the wave climate of the southeastern Australian coast. *Journal of Geophysical Research: Oceans*, 117(C11). DOI: 10.1029/2012JC008170.

Ranasinghe, R., Stive, M. J. F., and Smith, D. J. (2013). The effect of climate change on the dynamics of sandy beaches: A review. *Journal of Coastal Research*, 29(3), 355–367. DOI: 10.2112/JCOASTRES-D-12-00132.1.

Rock Manual. (2007). The use of rock in hydraulic engineering (2nd ed.). C683. CIRIA, CUR, CETMEF: London.

Rodriguez-Delgado, J., Muñoz, J. F., and García, M. (2020). Impacts of sea level rise on coastal flooding: A regional study. *Coastal Engineering*, 160, 103769. DOI: 10.1016/j.coastaleng.2020.103769.1.

Sardella, R., Davies, T., and Lowe, D. (2020). Climate change and coastal storm impacts: An analysis of future scenarios. *Environmental Research Letters*, 15(7), 074029. DOI: 10.1088/1748-9326/ab8f77.

Seabrook, S. R. and Hall, K. R. (1998). Wave Transmission at Submerged Rubblemound Breakwaters. https://doi.org/10.1061/9780784404119.150.

Sigurdarson, S. and Van der Meer, J. W. (2012). Wave Overtopping at Berm Breakwaters in Line with EurOtop. *Proceedings of 33rd International Conference on Coastal Engineering*, Santander.

Sigurdarson, S. and Van der Meer, J. W. (2013). Design of berm breakwaters: Recession, overtopping and reflection. *Proceedings Coasts, Marine and Breakwaters*, 18–20 September 2013, Edinburgh.

Sindhu, S. and Shirlal, K. G. (2015). Prediction of wave transmission characteristics at submerged reef breakwater. *Procedia Engineering*, 116, 262–268. https://doi.org/10.1016/j.proeng.2015.08.289.

Soares, C. G., Ramos, S., and Costa, M. (2017). The influence of climate change on coastal flooding and erosion: A review. *Coastal Engineering*, 124, 19–31. DOI: 10.1016/j.coastaleng.2017.03.003.

Strandberg, G., Barring, L., Hansson, U., Jansson, C., Jones, C., Kjellström, E., Kolax, M., Kupiainen, M., Nikulin, G., Samuelsson, P., Ullestig, A., and Wang S. (2014). CORDEX scenarios for Europe from the Rossby Centre regional climate model RCA4. *Reports Meteorology and Climatology*, 116, 1–84. https://www.smhi.se/polopoly_fs/1.90273!/Menu/general/extGroup/attachmentColHold/mainColl/file/RMK_116.pdf.

Suh, K. D., Kim, S. W., Mori, N., and Mase, H. (2012). Effect of climate change on performance-based design of caisson breakwaters. ASCE, *Journal of Waterways, Coastal and Ocean Engineering*, 138, 3, 215–225.

Takahashi, S., Shimosako, K., and Hanzawa, M. (2015). Design of coastal structures and sea defenses. *World Scientific*, 77–104.

Takahashi, S., Shimosako, K., Kimura, K., and Suzuki, K. (2000). Typical failure of composite breakwaters in Japan. *Proceedings 27th International Conference Coastal Engineering, ASCE, Reston*, VA, 1899–1910.

Tomasicchio, G. R. and D'Alessandro, F. (2013). Wave energy transmission through and over low crested breakwaters. *Journal of Coastal Research*, 65, 398–403. https://doi.org/10.2112/SI65-068.1.

Tomasicchio, G. R., Lamberti, A., and Guiducci, F. (1994). Stone movement on a reshaped profile. *Proceedings 24th International Conference on Coastal Engineering*, ASCE, Kobe, 2, 1625–1640.

Tørum, A., Kuhnen, F., and Menze, A. (2003). On berm breakwaters. Stability, scour, overtopping. *Coastal Engineering*, 49, 209–238 (Elsevier).

Tørum, A. and Sigurdarson, S. (2014). PIANC Working Group No. 40: Guidelines for the Design and Construction of Berm Breakwaters. Breakwaters, Coastal Structures and Coastlines, September 2001, London.

U.S. Army Corps of Engineers (CERC) (2003). Coastal Engineering Manual (CEM), Vol. 4.

UNEP. (2023). Global Environment Outlook 2023: Summary for Policy Makers. United Nations Environment Programme. DOI: 10.5281/zenodo. 7747942.

Van der Meer, J. W. (1988a). *Rock Slopes and Gravel Beaches under Wave Attack*, Doctoral Thesis. Delft University of Technology.

Van der Meer, J. W. (1988b). Deterministic and probabilistic design of breakwater armor layers. ASCE, *Journal of Waterways, Coastal and Ocean Engineering*, 114, 1, 66–80.

Van der Meer, J. W. (1990). Data on wave transmission due to overtopping. Delft Hydraulics. https://books.google.com.tr/books?id=Xni-tgAACAAJ.

Van der Meer, J. W., Briganti, R., Zanuttigh, B., and Wang, B. (2005). Wave transmission and reflection at low-crested structures: Design formulae, oblique wave attack and spectral change. *Coastal Engineering*, 52(10–11), 915–929. https://doi.org/10.1016/j.coastaleng.2005.09.005.

Van der Meer, J. W. and Daemen, I. F. (1994). Stability and wave transmission at low-crested rubble-mound structures. *Journal of Waterway, Port, Coastal, and Ocean Engineering ASCE*, 120(1), 1–19. https://doi.org/10.1061/(ASCE)0733-950X(1994)120:1(1).

Van der Meer, J. W. and Pilarczyk, K. W. (1990). Stability of low-crested and reef breakwaters. Delft Hydraulics report no. H986, prepared for CUR C67. https://doi.org/10.1061/9780872627765.105.

Van der Meer, J. W. and Sigurdarson, S. (2014). Geometrical design of berm breakwaters. *Coastal Engineering*, 34, 1–14.

Van der Meer, J. W. and Srigurdarson, S. (2016). *Design and Construction of Berm Type Breakwaters*, 40, 1st edition, World Scientific, Singapore.

Van Gent, M. R. A. (1995). Wave interaction with berm breakwaters. *Journal of Waterway, Port, Coastal, and Ocean Engineering*, ASCE, 121(5), 229–238.

Van Gent, M. R. A. (2013). Rock stability of rubble mound breakwaters with a berm. *Coastal Engineering*, 78, 35–45. https://doi.org/10.1016/j.coastaleng.2013.03.003.

Van Gent, M. R. A. (2019). Climate adaptation of coastal structures. *Keynote in Proceedings. Applied Coastal Research (SCACR 2019)*, Bari, Italy.

Van Gent, M. R. A., Buis, L., Van den Bos, J. P., and Wüthrich, D. (2023). Wave transmission at submerged coastal structures and artificial reefs. *Coastal Engineering*, 184. https://doi.org/10.1016/j.coastaleng.2023.104344.

Van Gent, M. R. A., Plate, S. E., Berendsen, E., Spaan, G. B. H., Van der Meer, J. W., and d'Angremond, K. (1999). Single-layer rubble mound breakwaters. *Proceedings Coastal Structures* 99, Santander, Spain, ASCE.

Van Gent, M. R. A. and Van der Werf, I. M. (2017). Single Layer Cubes in a Berm, SCACR 2017, 3–6 October 2017, Santander.

Vidal, C., Losada, M. A., and Mansard, E. P. (1995). Stability of low-crested rubble-mound breakwater heads. *Journal of Waterway, Port, Coastal, and Ocean Engineering*, ASCE, 121(2), 114–122.

Vidal, C., Losada, M. A., Medina, R., Mansard, E. P. D., and Gomez-Pina, G. (1992). A universal analysis for the stability of both low-crested and submerged breakwaters. *23rd International Conference on Coastal Engineering (ICCE) 1992*, Venice, Italy, 1679–1692. https://doi.org/10.1061/9780872629332.127.

Vitousek, S., Barnard, P. L., and Erickson, A. (2017). Coastal vulnerability to sea-level rise: Insights from a regional analysis of beach erosion and flooding. *Environmental Research Letters*, 12(9), 094013. DOI: 10.1088/1748-9326/aa8189.

Wang, S., Zhang, Y., and Chen, X. (2020). Assessing future coastal hazards using global climate models. *Ocean & Coastal Management*, 187, 105083. DOI: 10.1016/j.ocecoaman.2019.105083.

Yuksel, Y. (Ed.), (2016). *Manual of Planning and Design of Coastal Structures.* Ministry of Transport and Infrastructure, Ankara, Turkey.

Yuksel, Y., Cevik, E., Sahin, C., Van Gent, M. R. A., Gumus, S., Issever, D., Inal, U., and Ogur, U. (2025). Structural and hydraulic response of emerged low-crested cube-armoured breakwaters. *Applied Ocean Research (in progress).*

Yuksel, Y., Cevik, E., Van Gent, M., Sahin, C., Gulver, M., and Gultekin, C., (2018). Stability effects of cube armor unit placement configurations in the berm of a breakwater. *Proceedings of 36th International Conference on Coastal Engineering,* Baltimore.

Yuksel, Y., Cevik, E., Van Gent, M. R. A., Sahin, C., Altunsu, A., and Yuksel, Z. T. (2020). Stability of berm type breakwater with cube blocks in the lower slope and berm. *Ocean Engineering,* 217, 107985. https://doi.org/10.1016/j.oceaneng.2020.107985.

Yuksel, Y., Doran, B., Yuksel, Z. T., and Cevik, E. (2018). Seismic response of coastal and port structures. In Young, C. Kim (Ed.) *Handbook of Coastal and Ocean Engineering,* Expanded Edition, pp. 401–433. World Scientific.

Yuksel, Y., Van Gent, M. R. A., Cevik, E., Kaya, A. H., Ari Guner, H. A., Yuksel, Z. T., and Gumuscu, I. (2022). Stability of high density cube armoured breakwaters. *Ocean Engineering,* 253. https://doi.org/10.1016/j.oceaneng.2022.111317.

Yuksel, Y., Yuksel, Z. T., and Sahin, C., (2020). Effect of long-term wave climate variability on performance-based design of coastal structures. *Aquatic Ecosystem Health & Management,* 23(4), 407–416. DOI:10.1080/14634988.2020.1807302.

Zacharioudaki, A. and Reeve, D. E. (2011). Coastal response to climate change: A review of shoreline change and beach erosion. *Coastal Engineering,* 58(11), 969–987. DOI: 10.1016/j.coastaleng.2011.05.006.

Zanuttigh, B. and Van der Meer, J. W. (2008). Wave reflection from coastal structures in design conditions. *Coastal Engineering,* 55, 771–779. https://doi.org/10.1016/j.coastaleng.2008.02.009.

Zhang, S. and Li, X. (2014). Design formulas of transmission coefficients for permeable breakwaters. *Water Science and Engineering,* 7(4), 457–467. https://doi.org/10.3882/j.issn.1674-2370.2014.04.010.

Chapter 5

Storm Surge: Simulation, Impact, and Mitigation

S.A. Sannasiraj and Vallam Sundar

Department of Ocean Engineering,
Indian Institute of Technology Madras, Chennai, India

1. Introduction

Storm surge is defined as a rise in sea level caused by a storm, in particular, a tropical cyclone or hurricane. Storm surges, typically affecting low-lying coastal areas, result in severe flooding, loss of life, and significant property damage, and thus, these are some of the most critical concerns for coastal disaster management. The increase in the water level above the mean sea level can vary significantly, influenced by factors such as storm intensity, forward speed, coastline shape, and ocean floor slope. In some cases, surges can reach several meters, causing widespread flooding. Storm surges often are related to cyclone landfall. They can be amplified by astronomical tides, potentially causing catastrophic coastal flooding during the high tide. They can also force water upriver or into bays and estuaries, exacerbating flooding. Figure 1 depicts the surge on the landfall and subsequent landward inundation. The mechanisms driving storm surges involve a combination of atmospheric, oceanographic, and geophysical factors. These processes interact to generate and amplify the surge, leading to potentially catastrophic impacts on coastal areas.

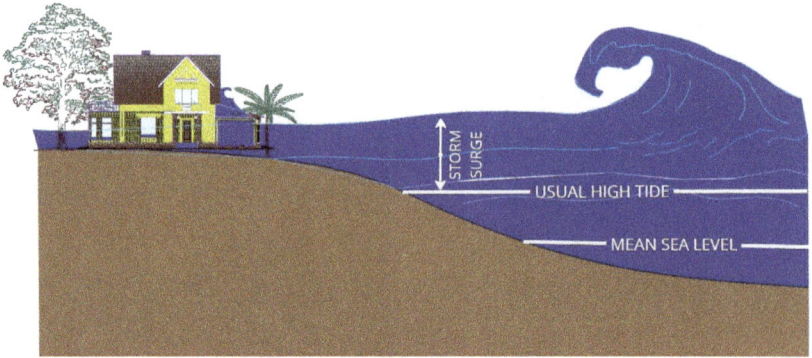

Fig. 1.   Representation of storm surge on landfall.

The key mechanisms driving storm surges are as follows:

Wind Stress: Wind stress is the strong winds of a tropical cyclone or a storm that induces surface stress over the ocean, pushing large volumes of seawater toward the coast. The wind transfers kinetic energy to the water, causing the surface layer to move in the direction of the wind. This phenomenon, known as wind-driven water transport, is the primary driver of storm surges. As water is driven ashore, it piles up along the coastline leading to significant rise in water levels and widespread inundation in low-lying coastal areas.

Pressure Gradient: The low-pressure center of a tropical cyclone reduces atmospheric pressure over the ocean, causing seawater to expand and rise. This mechanism contributes significantly to the total surge. Tropical cyclones are characterized by a central region of very low atmospheric pressure. The reduction in atmospheric pressure causes the seawater directly beneath the cyclone's center to expand vertically and it is observed as a localized rise in sea level. This is referred to as the inverted barometer effect. While it is of secondary importance compared to wind stress, the pressure gradient can add an additional 30–50cm to the surge height, especially in systems with extremely low central pressures.

Wind Waves: Cyclones produce large wind-driven waves, which accompany the storm surge and significantly exacerbate its impact. The energy from these waves combined with the surge causes coastal flooding and erosion. The waves can overtop seawalls and other coastal defenses, enhancing the risk of inundation and structural damage. Wave run-up and overtopping effects can extend the reach of floodwaters further toward inland.

Geophysical and Bathymetric Factors: The physical characteristics of the coastal region and the underwater topography play a critical role in determining the magnitude of a storm surge. For instance: (i) Funnel-shaped coastlines, which are regions with coastlines that narrow inward such as the Bay of Bengal, act to focus and intensify incoming water leading to the amplification of surge and (ii) shallow continental shelves which are areas with gently sloping seabed allow for greater accumulation of water as it approaches the coast, leading to higher surges. These factors can magnify the surge by several meters with devastating consequences for densely populated coastal areas.

Some of the key features of a storm surge are as follows:

Height: Storm surge can raise the sea level by several meters that increases the risk of coastal flooding. The height of the storm surge depends on several factors, such as the intensity of the storm, its forward speed, the shape and slope of the coast, and the tidal range.

Timing: Storm surge usually occurs when a storm makes landfall, as the strong winds push water toward the shore. It is also affected by astronomical tides, which can amplify the effects of the storm surge.

Coastal inundation: The most significant impact of storm surge is coastal inundation, which can cause severe damage to coastal communities. Floodwaters can penetrate far inland which can cause widespread damage to homes, businesses, and infrastructure.

Inland flooding: Storm surge can also cause flooding in rivers, bays, and estuaries as the increased water level can cause water to back up and overflow.

Dangers: In addition to the damage caused by the flooding, storm surge can also pose a threat to life and property due to the strong winds, high waves, and dangerous currents associated with the surge.

1.1 *Most devastating surges around the globe showed the devastating impact on coasts. Few examples are highlighted in the following:*

- **Hurricane Katrina**, one of the most devastating hurricanes with wind speeds of about 205 km/hr, produced a storm surge of about 8.5 m during its landfall on August 29, 2005. It had resulted in destruction in particular along the coast of Louisiana, Mississippi, and Alabama with over 1800 deaths and loss of property worth $125 billion and displaced about 1.5 million people. Despite warnings, many residents failed to evacuate due to mistrust, lack of resources, or underestimation of risk (Knabb *et al.*, 2005).

- **Super Typhoon Haiyan also termed as Typhoon Yolanda** struck the coast of Central Philippines in 2013 with wind speeds up to 315 km/hr and produced a storm surge of about 7.5 m along Tacloban City and adjoining coastal stretches. It demonstrated the destructive potential of intense tropical cyclones, emphasizing the need for robust disaster preparedness, improved early warning systems, and the restoration of natural coastal defenses. It remains a critical case study for understanding storm surge impacts in the context of climate change. It was officially reported that over 6,300 deaths and more than 4 million people were displaced with an estimated loss of about Estimated losses exceeded $5.8 billion (Perez *et al.*, 2016).

- **The Bay of Bengal** (BoB) located on the northeastern part of the Indian Ocean accounts for a significant portion of the world's tropical cyclones, with an average yearly frequency of 5–6% of global cyclones. The tropical cyclones that form over the Bay of Bengal during the pre-monsoon (April–May) and post-monsoon (October–November) are often intense, with wind speeds exceeding 150 km/h in severe cases. The low atmospheric pressure at the center of cyclones creates a "suction effect," pulling seawater upward, contributing to the surge. Cyclones in the bay are often slow-moving, allowing storm surges to gain strength and cause prolonged inundation. The semi-enclosed, funnel-shaped geography of the BoB directs cyclonic energy toward the coast, amplifying the height and intensity of storm surges. In addition, the shallow seabed in the northern Bay of Bengal amplifies storm surge heights, as there is less water depth to absorb the energy of incoming waves. The BoB is one of the world's most cyclone-prone areas experiencing cyclones approximately five times more frequently than on the western part of the Indian Ocean, i.e., on the Arabian Sea. Over 75% of TCs in the North Indian Ocean (NIO) occur here. The BoB experiences some of the world's highest storm surges, with heights exceeding 13 meters during severe events. These surges impact the low-lying coastal areas particularly vulnerable to high surge levels along the northeastern coasts of India and Bangladesh. The eastern Indian coastline along the BoB which is home to over 250 million people faces severe risk due to its densely populated and low-lying terrain. Major rivers, including the Mahanadi, Krishna, and Godavari, increase susceptibility to inland flooding during surges, as they provide pathways for seawater intrusion (Lal and Harasawa, 2001).

1.2 *Coastal Impacts*

The significant impact of the surge on the coastal region can be broadly categorized into human losses and infrastructure and ecosystem damages. In BoB, the historical storm surges have caused catastrophic casualties, such as the following: over 300,000 fatalities during the 1970 Bhola Cyclone (Bangladesh), thousands of deaths and widespread devastation during the 1999 Odisha Super Cyclone (India), and the economic losses exceeding $20 billion during the 2020 Cyclone Amphan. The coastal infrastructure, agriculture, and ecosystems (mangroves, coral reefs) face destruction due to inundation and erosion. It is important for the coastal communities to be prepared for a storm surge or a similar or even more devastating coastal hazard such as a tsunami, by knowing the evacuation routes, having an emergency plan, and staying informed about weather conditions during a storm. With proper preparation, the impacts of storm surges can be minimized, protecting lives and property. Computer models are used to simulate the physical processes of storms and their associated surges. These models use data from satellite observations, weather forecasts, and other sources to predict the height, timing, and landfall location.

2. Prediction of Storm Surges

The prediction of storm surges relies on real-time data collection, analysis, and the application of sophisticated numerical models, which are discussed below.

2.1 Data sources

Tide gauges are instrumental in measuring sea-level changes in real time providing crucial data on both normal tidal variations and abnormalities created by storms which depends on its intensity. Measured data from multiple locations help in mapping the extent and intensity of the surge along the coastline. Rapid rises in sea level detected by tide gauges can indicate the onset of a storm surge. Weather stations contribute atmospheric data, including wind speed, direction, barometric pressure, temperature, and precipitation. These measurements, in particular the barometric pressure and wind dynamics, are vital in understanding the intensity and trajectory of the storm, which are key determinants for the behavior of storm surge. A sharp drop in atmospheric pressure recorded at a weather station signals the approach of the cyclone's low-pressure core. This would also serve as an initial warning and following it track would help mitigate against the impacts along the coast.

Satellite observations provide a broader spatial perspective, capturing data on sea surface temperatures, wave characteristics, as well as atmospheric conditions; whereas, Drones and Lidar can provide aerial observations of coastal impacts, while buoys anchored in the ocean measure wave heights, sea temperatures, and currents. Remote sensing is handy in understanding the status of the shoreline to investigate the post coastal hazard events.

2.2 Forecast models

Forecasting numerical models if properly modeled can simulate the behavior of storm surges using mathematical equations considering the physical processes such as wind stress, pressure gradients, and wave dynamics for the given coastal topography. The observations from tide gauges, weather stations, and satellites would steer the model for better prediction. These include current sea levels, wind speeds, pressure fields, and forecasted storm paths. The models predict the surge, its extent, and the timing of the storm surge at specific locations.

2.3 Types of models

Hydrodynamic models are computational tools adopted for the simulation and analysis of the motion and behavior of water in oceans, rivers, lakes, and estuaries. The prediction through these models on the motion of water and its interaction with the surrounding elements such as structures and bathymetry is achieved by solving mathematical equations that govern fluid motion, primarily based on the principles of physics such as the conservation momentum using the Navier–Stokes equations. The external forces that are to be considered are the wind stress, pressure gradient, tidal variation, sources, and sinks such as river inflow and outflow. The models incorporate key hydrodynamic processes such as advection and diffusion, stratification, as well as wave propagation and deformation that may be due to temperature or salinity gradients. The hydrodynamic models are widely applied for flood prediction, water quality management, coastal engineering, sediment dynamics, climate change investigations, the fate of effluent disposal in coastal waters, etc. Probabilistic models are also adopted to account for the uncertainties by generating multiple scenarios based on the variations in storm characteristics and environmental conditions. The factors that are to be considered for predictions are current conditions, weather patterns, geographical and bathymetric data, and historical data. Accurate prediction allows meteorological agencies to issue storm surge warnings well in advance, enabling timely evacuations and preparations. The prediction has to be accurate so as to facilitate risk assessment and mitigation. A false alarm has to be avoided in total.

3. Numerical Simulation of Storm Surges

3.1 *General*

Mathematical equations that represent environmental processes are used to create simulations through numerical approximations over a divided time and space domain. However, accurately predicting the hydrodynamic and sediment dynamics at a specific location through numerical models can be challenging due to the numerous complex interactions involved (Chen and Cai, 2018; Wang and Li, 2017). Numerical models make some assumptions for calculating these interactions. The errors arising from such assumptions are reduced by calibrating and validating the results with data collected from the site.

Accurate modeling of storm surges, which are a significant impact of cyclones along coastal cities, requires a proper representation of the cyclonic wind fields in order to fully capture the unique characteristics of a particular storm event (Bloemendaal *et al.*, 2019). The stretches along the east coast of India are more prone to high water surface elevation due to extreme events. The wind induces waves on the ocean surface and it is evident that stronger storms generate higher waves (Glejin *et al.*, 2013).

The holistic representation of the numerical modeling approach simulates the interaction of surges and waves, and further, it can be integrated into morpho dynamics evolution and inundation mapping. The high accuracy of the surge estimate can be ensured with the enhanced wind fields and coupling by reducing uncertainties in predictions. The numerical tool is scalable to different coastal regions and storm intensities. It can assess future impacts of climate change on storm surges and coastal morpho dynamics.

The hydrodynamic modeling simulates the movement of water, including storm surges and tides, by solving the shallow water equations (SWEs). The SWEs account for the conservation of mass and momentum in two dimensions. It incorporates wind forces acting on the water surface, a critical factor during cyclones, and thus, it simulates the effects of low atmospheric pressure from cyclones. The influence of astronomical tides interacts with storm surges. The energy dissipation due to seabed friction, especially significant in shallow coastal areas, can be accounted for.

The wave modeling accounts for the generation, propagation, and dissipation of wind-generated waves and their interaction with coastal dynamics. It provides the additional water-level rise due to wave action near the shore. The simulation of wave energy dissipation as waves approach shallow waters provides a better estimate of sediment transport in addition to surge. The wave–current interaction accounts for response mechanisms where currents affect the wave behavior and vice versa. The storm-wave influence captures the intensification of waves during cyclonic events, which contributes to coastal flooding. The wave models thus help predict coastal inundation and erosion, providing insights into how waves amplify storm surge impacts. The dynamic coupling of hydrodynamic and wave models would ensure accurate predictions of surge. Waves influence water levels and currents through radiation stress and the water flow modifies wave propagation patterns.

Combined wind fields, such as GAHM and WRF, enhance the simulation of nearshore dynamics, improving our understanding of coastal processes. Coupled hydrodynamic-wave-morpho dynamic frameworks are particularly effective in predicting water levels, wave heights, and sediment changes. Tools such as TCSPI provide reliable surge predictions and hazard mapping and help in the preparation of evacuation plans and for mitigation against the impacts of severe weather events. The integration of machine learning algorithms helps refine models using large datasets from past events. Higher-resolution climate and ocean models improve spatial and temporal predictions. By leveraging the synergy between observational data and sophisticated forecast models, the prediction of storm surges becomes a powerful tool in safeguarding coastal communities against the devastating impacts of extreme weather events.

The machine learning (ML) models demonstrate high predictive accuracy and computational efficiency, making them valuable tools for analyzing and forecasting coastal hazards. It is expected that surge heights will increase by 10–18% due to rise in the sea levels and more intense and frequent cyclones, highlighting the importance of continued advancements in predictive modeling and coastal management strategies.

3.2 *Advancement in the prediction of storm surge*

Accurate prediction and modeling integrate numerical modeling, statistical approaches, and machine learning to improve predictive accuracy and computational efficiency. The approach should particularly be tailored to capture the intricate dynamics along the coast, where tropical cyclones and storm surges are frequent and impactful.

Statistical models use historical data to predict surge heights based on past cyclone characteristics. Machine learning (ML) bridges the gap between numerical and statistical models, offering rapid predictions with enhanced accuracy.

4. Coastal Flooding and Mitigation

4.1 *Flooding*

Coastal flooding, increasingly linked to climate change, poses significant threats to coastal communities worldwide. This phenomenon arises from a combination of natural processes and human activities, making it a critical area of study in environmental science, urban planning, and policy-making. To predict and understand coastal flooding, various simulation techniques are employed. Hydrodynamic models, such as Delft3D (Deltares, 2011) and MIKE 21 (DHI, 2021), use fluid dynamics equations to simulate water movement and predict flood scenarios. Storm surge models, such as the SLOSH (Sea, Lake, and Overland Surges from Hurricanes) model of Jelesnianski *et al.* (1992), integrate atmospheric data to focus specifically on surge predictions. Statistical models rely on historical data to estimate flooding probabilities under diverse conditions, while climate models project long-term sea-level rise and its implications. Emerging machine learning techniques enhance these approaches by leveraging big data to improve accuracy in flood prediction and risk assessment.

The numerical model based on finite element method, ADCIRC (Luettich and Westerink, 1992), to simulate water levels and currents over large domains with high resolution is widely used for storm surge modeling and coastal circulation analysis.

4.2 *Mitigation measures*

4.2.1 *General*

Mitigation strategies for coastal flooding encompass structural and non-structural measures. Structural solutions include constructing seawalls, levees, and storm surge barriers to protect vulnerable areas. Beach nourishment and dune restoration are effective in buffering wave energy and preventing erosion. Developing resilient infrastructure ensures that buildings and transportation networks can withstand flooding events. Non-structural measures, such as implementing early warning systems and adopting land-use planning to avoid high-risk areas, play a crucial role in reducing exposure to flooding. Conserving wetlands and promoting sustainable coastal development help create natural buffers against storm surges and rising seas. Community-based approaches, including public education on flood risks and preparedness, empower individuals and local groups to participate in disaster response planning. The readers may refer to chapter 4 of Sundar and Sannasiraj (2019). Innovative solutions are increasingly important in addressing coastal flooding. Green infrastructure, such as mangroves and oyster reefs, provides natural protection against waves while enhancing biodiversity. Smart technologies enable real-time monitoring and adaptive management, allowing for rapid response and efficient resource allocation during flooding events.

The Netherlands is renowned for its advanced flood defense systems, including the Delta Works project, Jos Van Alphen (2015), which incorporates barriers, sluices, and levees to protect low-lying areas. In Bangladesh, a combination of community-based adaptation and structural measures addresses frequent flooding in one of the most vulnerable regions globally.

Following Hurricane Sandy, New York City implemented the "Big U" project (Fukuoka *et al.*, 2021) an ambitious initiative to create a resilient urban coastline through integrated flood protection systems and community engagement. Looking to the future, addressing coastal flooding will require continued research and innovation. Enhancing the integration of climate models with local-scale flood simulations can improve prediction accuracy and inform targeted interventions. The effective measures for flood control are discussed in the following section.

4.2.2 *Structural countermeasures*

4.2.2.1 General

Structural countermeasures are engineered solutions designed to reduce the impacts of flooding on coastal areas. These measures are typically categorized into three types: blocking, steering, and slowing, as illustrated in Fig. 2. Blocking measures, such as seawalls, dikes, and offshore breakwaters, aim to prevent surges from reaching inland. These structures are particularly effective in protecting densely populated or economically vital regions but come with challenges such as high construction costs, potential failure during extreme events (e.g., overtopping), and the risk of creating a false sense of security. Steering measures, such as flood channels, berms, and topographic depressions, are designed to direct the flow away from vulnerable areas or infrastructure. These solutions redistribute the kinetic energy, minimizing damage in critical zones. However, they require careful planning and significant land use to avoid unintended amplification of damage in adjacent areas. Slowing measures and including vegetation, buffer blocks, and porous walls work to dissipate the speed of flow and reduce inundation depth and velocity. These methods are especially useful in densely developed regions where space is limited. To enhance effectiveness, integrated and innovative approaches are increasingly emphasized. Multi-layer safety systems combine blocking, steering, and slowing measures with emergency planning and spatial management, offering a comprehensive defense strategy. Emerging technologies such as recurved seawalls and submerged barriers, as well as reinforced natural defenses such as mangroves, add further resilience. Tailoring these

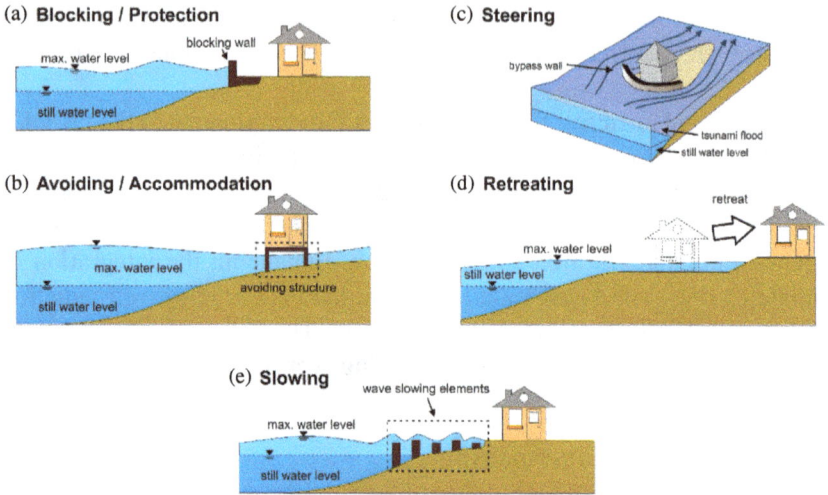

Fig. 2.   Basic strategies to reduce Tsunami risk following NOAA (2001, reworked).

structural measures to regional tsunami risks, socioeconomic conditions, and environmental considerations is critical for achieving optimal protection.

4.3 *Submerged Breakwaters (SBWs)*

Submerged breakwaters (SBWs) are offshore structures designed to dissipate wave energy and provide coastal protection. The **smooth** and **stepped trapezoidal SBWs**, for their hydrodynamic performance in wave attenuation, are considered here.

Smooth submerged breakwaters are designed with a continuous, uniform front slope, which allows waves to slide up the smooth surface, leading to partial reflection and energy dissipation. Hydrodynamically, they exhibit lower energy dissipation compared to stepped designs and are less effective in inducing premature wave breaking. The advantages of smooth submerged breakwaters include their simple design and construction, as well as lower construction costs compared to stepped designs. A comprehensive review of the literature on the behavior of submerged reefs is reported by Lokesha *et al.* (2013).

Stepped submerged breakwaters are designed with front slopes segmented into steps. These steps are specifically engineered to create turbulence and induce wave breaking earlier than smooth designs. Hydrodynamically, they offer higher wave energy dissipation due to the enhanced turbulence and wave breaking, making them more effective at reducing the wave transmission coefficient (Kt). The advantages of stepped submerged breakwaters include increased efficiency in wave attenuation and effectiveness in both regular and irregular wave conditions. For details, reader may refer to Lokesha *et al.* (2015).

Stepped and smooth sloping submerged breakwaters (SBWs) as shown in Fig. 3 dissipate more wave energy compared to smooth ones, especially in high-energy wave environments. The stepped design promotes earlier wave breaking, enhancing tranquillity on the lee side of the breakwater. The relative crest width (B/L, where B is the crest width and L is the wavelength) significantly influences wave transmission. Optimal B/L ranges (e.g., 0.6–0.7) have been identified for stepped SBWs to maximize energy dissipation. Practically, stepped SBWs are more suited for locations with high wave energy or where wave attenuation is critical, whereas smooth SBWs can be used in calmer wave environments or where budget constraints exist.

Field applications of submerged breakwaters (SBWs) provide critical insights into designing and deploying these structures for real-world coastal scenarios. For coastal protection, the primary objective is to reduce wave energy reaching the shore to prevent coastal erosion. Stepped SBWs are typically placed offshore in high-energy wave environments, while smooth SBWs are more suitable for calm or moderately energetic coastal areas. To maintain harbor tranquillity, SBWs are strategically positioned at harbor entrances to reduce wave penetration, with stepped SBWs featuring optimal crest width and submergence ratios offering superior tranquillity. In terms of wave energy management, submerged breakwaters dissipate wave energy, allowing controlled wave transmission for activities such as surfing. Additionally, environmental integration of

Fig. 3. A view of smooth and stepped trapezoidal submerged breakwaters (SBWs).

SBWs enhances water circulation and sediment exchange between off-shore and nearshore areas, reducing beach pollution and improving marine habitats. Key design recommendations highlight the importance of placement distance from the shore to ensure effective wave attenuation without negatively affecting sediment transport. For extended coastlines, deploying a series of SBWs with optimized gaps can balance wave dissipation and water exchange.

4.4 *Submerged trapezoidal artificial reef units (STARUs)*

Submerged Trapezoidal Artificial Reef Units (STARUs), as shown in Fig. 4, are engineered structures designed to attenuate wave energy, reduce wave transmission, and promote marine biodiversity. These structures come in two design variants: STARU-0, which are non-perforated solid units that reflect and dissipate wave energy, and STARU-11, which are perforated units with 11% perforation, allowing partial flow-through to enhance turbulence and reduce structural stresses. STARU's modular configuration includes single, double, four, and eight-unit setups, studied to evaluate the effects of unit numbers and trench width on wave attenuation (Lokesha *et al.*, 2019). Multiple units are arranged with gaps or

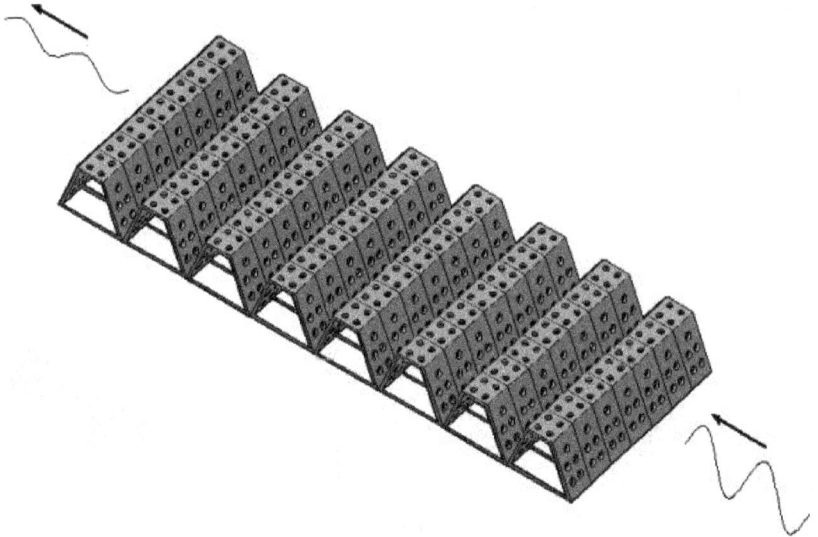

Fig. 4.   Multiple submerged trapezoidal artificial reef units (STARUs).

trenches to optimize energy dissipation. In terms of hydrodynamic behavior, STARU effectively reduces the wave transmission coefficient (Kt) by dissipating energy through turbulence, wave breaking, and dynamic interactions between reef units. Notably, perforated units demonstrate improved wave dissipation and lower dynamic pressures compared to their non-perforated counterparts.

The performance analysis of STARUs reveals key insights for both single- and multiple-unit configurations. For single units, STARUs are most effective in localized areas or where space is limited. The perforated STARU outperforms the non-perforated system under turbulent flow conditions, due to its flow-through mechanisms. When multiple units are used, wave attenuation is significantly improved. This is achieved by increasing the overall width of the structure and enhancing energy dissipation through successive wave interactions. A critical factor in the performance of multiple units is the optimal trench width, as wider trenches allow more turbulence, leading to an efficient reduction of wave energy. Several key parameters influence performance: the relative crest width (B/L), where wave transmission is minimized in the range B/L≈0.5–0.7; the degree of submergence (d/h), where submergence ratios beyond d/h = 1.2 yield diminishing returns; and the trench width (Bt/B), where an optimal trench width enhances energy dissipation without compromising the structural stability.

The STARUs have various applications across different fields, showcasing their versatility and effectiveness. In coastal erosion prevention, STARUs act as submerged breakwaters, reducing wave energy before it reaches the shoreline. They help prevent beach erosion by minimizing wave action while allowing for sediment transport and natural water circulation. For harbor tranquillity, STARUs are deployed at harbor entrances or within approach channels to reduce wave heights and ensure calm conditions for navigation and berthing. Multiple STARUs can provide uniform wave attenuation across large areas, enhancing safety and efficiency. Additionally, STARUs contribute to wave energy management, protecting offshore infrastructure and enabling aquaculture farming. Designed for climate resilience, they adapt to rising sea levels with optimized submergence ratios, offering sustainable alternatives to traditional seawalls. These diverse applications illustrate the importance and benefits of STARUs in various coastal and marine environments.

STARUs were deployed to counter coastal erosion and rehabilitate corals along the Vaan Island, one of the vulnerable stretches within the

Gulf of Mannar. The trapezoidal reef configurations successfully reduced wave energy and promoted marine ecosystem recovery.

4.4.1 *Channels and dug pools*

Channels and dug pools are innovative flood control measures that act as natural or engineered buffers to reduce the impact of storm surges and also tsunamis. These features work by absorbing the energy of incoming waves, slowing their velocity, and regulating their flow before they reach critical inland areas. During the 2004 Indian Ocean Tsunami, the Buckingham Canal in Chennai, India, provided a striking example of this approach's effectiveness. The 30-meter-wide and 10-meter-deep canal, situated about 1–2 km from the shoreline, served as a buffer zone, significantly reducing tsunami run-up and channeling the overflow back to the sea within minutes. This natural occurrence highlighted the potential for similar structures to be integrated as deliberate tsunami countermeasures.

Experimental and numerical studies have further validated the effectiveness of channels and dug pools. Research indicates that wider and deeper canals perform better in dissipating wave energy, although they do not significantly reduce inundation depth on their own. When combined with other measures, such as sand dunes or vegetation belts, these features demonstrate a marked reduction in both flow velocity and inundation levels. For instance, studies on the Kita-Teizan Canal in Sendai, Japan (Mohammad and Tanaka, 2016), suggest that such channels can reduce tsunami flow velocity by 13–20%, with corresponding decreases in structural damage in protected areas.

Elevated structures are vital for tsunami mitigation, serving as barriers to reduce inundation and as evacuation points when natural high ground is unavailable. Examples include elevated roads, embankments, reinforced buildings, and vertical evacuation shelters, often integrated into existing infrastructure for cost efficiency. These structures are especially effective in urban areas with limited evacuation time and space. Atlantic City is the birthplace of boardwalks in America. It opened on June 26, 1870, introducing the first removable wooden walkway, a view of which is shown in Fig. 5.

With proper planning, such as adequate height, strategic placement, and robust structural design, elevated structures can withstand tsunami forces and enhance safety. Challenges include high costs and ensuring

Fig. 5.   A view of the boardwalk in Atlantic City.

structural integrity during extreme events. However, repurposing existing infrastructure provides a feasible alternative in resource-constrained regions. **By integrating elevated structures or dug channels with other measures such as seawalls, channels, and warning systems, coastal resilience can be significantly strengthened.**

4.5. *Slowing by artificial elements*

Slowing by artificial elements is an innovative approach to tsunami mitigation that focuses on reducing the inundation distance and time by dissipating wave energy along the coastal stretches. These artificial elements act similarly to natural vegetation, creating turbulence and dissipating energy as waves pass through and over them. Examples include buffer blocks, perforated seawalls, and other macro-roughness elements designed to attenuate the wave's energy before it reaches critical infrastructure. These structures are particularly effective in densely populated regions where space for conventional measures such as wide breakwaters or dikes is limited.

Buffer blocks, for instance, have been widely studied for their potential in mitigating high-energy waves. The idea of buffer blocks as a mitigation measure has originated from the ones down the spillway for flow attenuation (Fig. 6).

(a)   (b)

Fig. 6.   Buffer blocks (a) down a spillway and (b) in the Island of Nordeney, Germany.

Originally used against storm surges, buffer blocks and perforated seawalls are gaining attention for tsunami mitigation due to their space-saving design and energy-dissipation capabilities. Staggered buffer block configurations and variations in shape (e.g., rectangular or semi-circular) optimize performance, while perforated seawalls reduce wave height and impact forces by up to 40% and material costs by 25%. Challenges include the need for precise design, limited field data, and the lack of standardized guidelines. Research has largely been small-scale, with large-scale applications still underexplored. Combining slowing elements with natural defenses, such as vegetation belts, or integrating them into existing infrastructure offers a cost-effective, adaptable strategy to enhance coastal resilience against tsunamis.

4.5.1 *Multi-layer approach*

The multi-layer approach integrates structural measures prevention (Layer 1), **spatial planning** (Layer 2), and **emergency management** (Layer 3) to minimize damage and loss of life. It ensures redundancy, with subsequent layers providing protection if one fails. Examples include seawalls, land-use planning, and early warning systems, tailored to regional economic and geographic conditions. While developed nations often implement all three layers, cost-effective combinations, such as

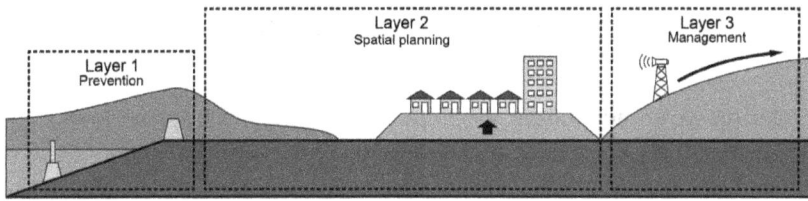

Fig. 7. A schematic view of the multi-layer approach. Layer 1: Prevention (e.g., by offshore breakwaters or seawalls). Layer 2: Spatial planning (e.g., creating retention areas or lifted structures with porous structures). Layer 3: Management (e.g., evacuation plans and early warning systems). (Redrawn and extended from Tsimopoulou *et al.*, 2012, 2015).

elevated roads with evacuation drills, can enhance resilience in developing regions. Notable applications include the Dutch flood safety model and adaptations in Japan, where gaps in emergency planning during the 2011 Tohoku tsunami underscored the need for better integration. Challenges include coordination among stakeholders, dependency between layers, and ensuring effective design and public awareness. By combining structural defenses with proactive planning and emergency management, the multi-layer approach offers a robust, adaptable framework for reducing tsunami risks globally. A typical schematic view of the multi-layer defense approach is shown in Fig. 7.

5. Vegetation-Based Adaptation

5.1 *General*

5.1.1 *Coastal regions and their geomorphological features*

Coastal regions are dynamic ecosystems shaped by land, sea, and atmospheric interactions. They support biodiversity, provide resources, and act as buffers against natural disasters. Sandy beaches and dunes protect against wave action and erosion, with vegetation enhancing dune stability. Estuaries and lagoons absorb storm surges, reduce flooding, and host unique biodiversity. Mangroves and tidal flats prevent erosion, trap sediments, and support nutrient cycling and migratory birds. Coastal forests, such as Casuarina and Pandanus, reduce wave energy during extreme events. However, these vital features face threats from climate change and human activities, emphasizing the need for sustainable management.

5.1.1.1 The role of vegetation-based adaptation in coastal management

Vegetation-based adaptation uses coastal ecosystems to provide sustainable protection, complementing natural processes. Creepers such as *Ipomoea pes-caprae* stabilize sand dunes, reducing erosion and promoting sand accretion. Mangroves, with intricate roots, reduce wave energy, trap sediments, and mitigate storm surges while aiding land elevation. Coastal forests act as windbreakers, shielding inland areas from cyclonic winds. Specialized vegetation, such as tsunami-resistant Casuarina and Pandanus, absorbs wave energy and reduces inundation. This approach enhances natural defenses, supports biodiversity, and improves water quality, offering ecological and protective benefits. The role of vegetation in reducing the inundation distance and height was clearly demonstrated during the tsunami in 2004, as can be seen in Fig. 8.

Impacts of Vegetation-Based Adaptation in Mitigating Cyclonic Impacts
Cyclonic events are among the most destructive natural disasters affecting coastal regions, leading to severe erosion, flooding, and loss of property and life. Vegetation-based adaptation has proven to be a highly effective

Fig. 8.   Role of vegetation during extreme coastal hazards.

strategy in mitigating the impacts of such events. During Cyclone Gaja in 2018, the presence of creepers on sand dunes along the Karaikal coast (10.9254° N, 79.8380° E) played a critical role in minimizing erosion. While the above-ground vegetation was damaged, the dunes remained intact, preventing extensive inland flooding and showcasing the resilience provided by vegetation.

Mangroves are effective buffers against cyclones, dissipating wave energy, trapping sediments, and reducing storm surge impacts. During Cyclone Amphan, mangrove-covered areas in India and Bangladesh suffered less damage, emphasizing the need for their restoration. Similarly, tsunami vegetation such as Casuarina proved effective during Cyclone Nivar, with dense vegetation minimizing erosion compared to open or seawall-protected areas. Integrating vegetation with sand dunes enhances coastal resilience, as dunes stabilize the coast and vegetation mitigates wave impact. Beyond protection, these systems aid in faster recovery and restore coastal stability post-cyclone.

5.2 Case studies

GAJA cyclone

Karaikal's coastline is distinguished by unique geomorphological features and well-planned infrastructure, with abundant coastal vegetation that prevents erosion and supports ecological stability. Sand dunes, stabilized by native creepers and plants, act as natural barriers against coastal erosion, forming a key part of the region's defense system. Cyclone Gaja made landfall on November 16, 2018, with 120 km/h winds. Pre-cyclone data were collected on November 8 and post-cyclone measurements on November 17. Beach profiles and storm surge data were gathered at seven strategically chosen locations to assess coastal dynamics. Five transects were located north of the training walls and two to the south. Transects 1 and 2 were along an open beach, 3 and 4 between sand dunes and creepers, 5 on a dune with creepers, and 6 and 7 in front of vegetation cover. This approach provided insights into coastal responses to the cyclone, highlighting the role of natural features and structures in mitigating its impacts. The details of the study area along with the transects are shown in Fig. 9.

The coastal changes observed over four distinct periods — pre-cyclone data from October 2018, post-cyclone data collected immediately after the cyclone, one month post-cyclone, and one year post-cyclone — are

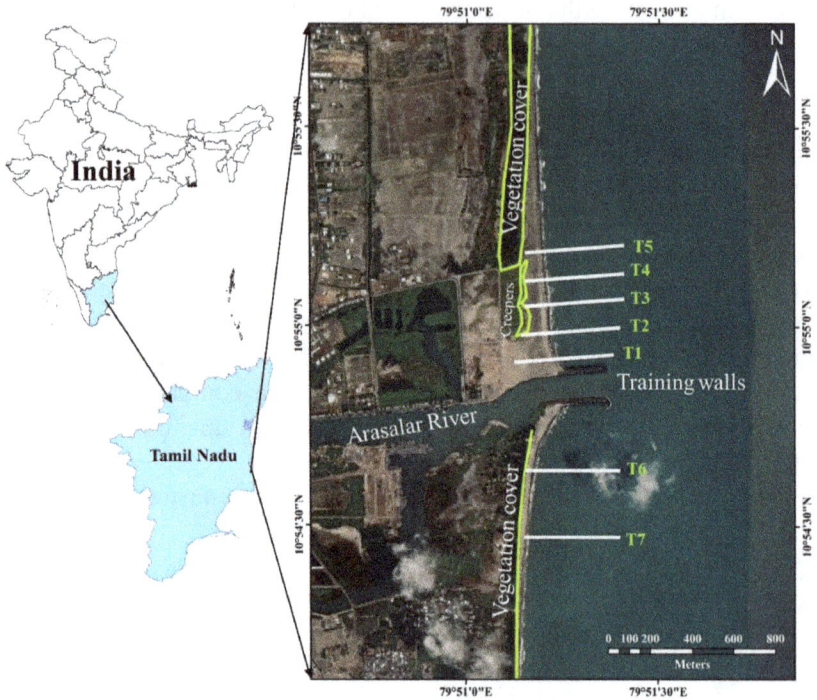

Fig. 9.   Study area with site description and transect locations (Transect 1 to Transect 7).

analyzed. These temporal comparisons provide a comprehensive under-standing of the immediate and long-term impacts of the cyclone on beach profiles, sand dunes, vegetation cover, and sediment dynamics, highlight-ing both recovery processes and persistent changes.

The analysis of shoreline changes (10) in Karaikal provides critical insights into the impacts of the Very Severe Cyclonic Storm (VSCS) Gaja and the subsequent recovery dynamics. The transect-wise data from October 2018 (pre-cyclone), immediately after the cyclone, December 2018 (one month post-cyclone), and October 2019 (one year post-cyclone) reveal significant temporal variations in the shoreline positions.

5.2.1 *Immediate impact of the cyclone*

The Gaja Cyclone caused notable landward displacement of the shoreline across all transects, indicating severe erosion. The cyclone's high wind

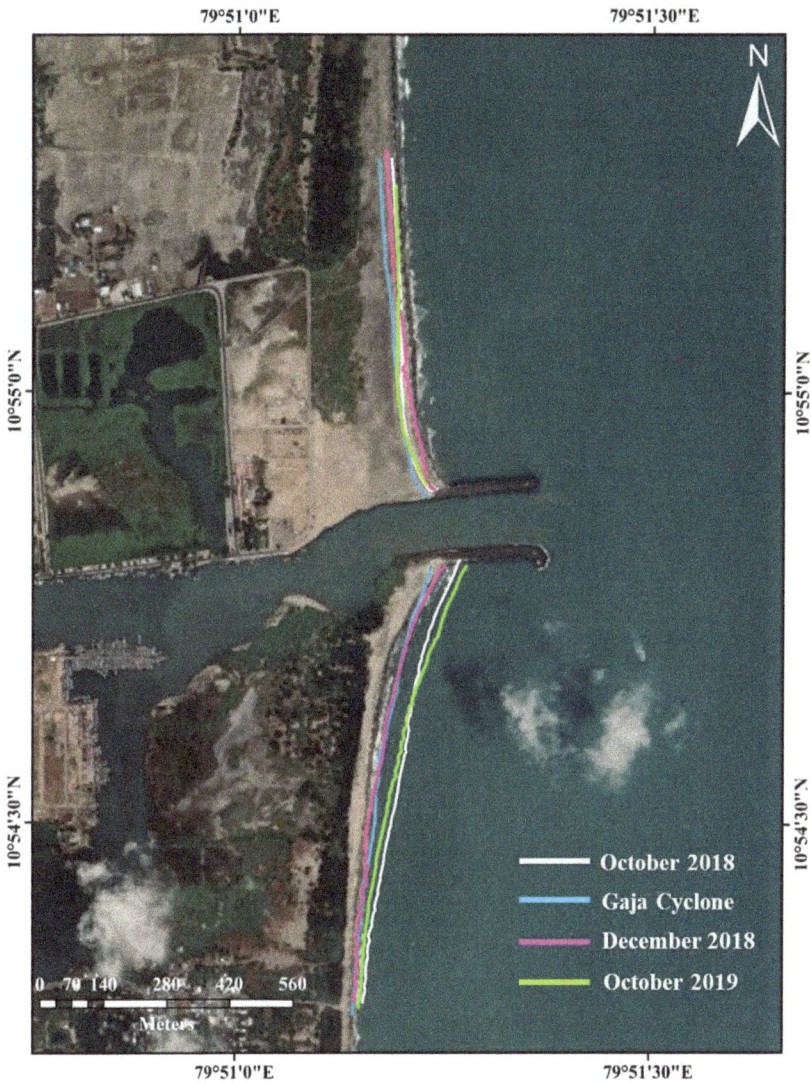

Fig. 10. Shorelines comparison for Karaikal.

speeds, storm surges, and wave actions likely intensified sediment transport, destabilizing the coastline. Transects located closer to open beaches,
such as Transects 1 and 2, experienced the most significant impacts, with
shoreline positions shifting landward by approximately 30–40 meters in

some cases. This immediate impact exposes the vulnerability of the areas to high-energy events.

The post-cyclone data from December 2018 shows that the shoreline began to recover within a month. Sediment deposition and reduced wave energy during the recovery period contributed to this stabilization. By October 2019, most transects had nearly regained their pre-cyclone positions, reflecting the coastline's natural resilience. Transects 5, 6, and 7, which are located closer to vegetated areas, exhibited faster and more consistent recovery compared to the more exposed transects (Transects 1 and 2).

5.2.2 Beach profile changes

The beach profile changes in the Karaikal region were analyzed across seven transects to assess the impacts of the Gaja Cyclone and subsequent recovery processes. These profiles (11), plotted in the following, provide a detailed representation of the erosion, sediment redistribution, and recovery dynamics observed over different time periods.

The beach profile changes in Karaikal, analyzed across seven transects, show the Gaja Cyclone's significant impact and recovery over a year. The cyclone caused erosion and sediment redistribution, with variations based on local geomorphic features and natural-engineered interactions. Areas with dense vegetation, particularly creepers on sand dunes, experienced minimal changes. These plants effectively shielded the dunes from erosion, acting as a natural buffer against the cyclone's wave forces.

However, the severe winds and intense wave action during the cyclone caused the destruction of the creeper cover. Despite this loss, the sand dunes underneath remained largely intact and were protected from erosion, showcasing the resilience provided by the vegetative cover. The vegetation served as the first line of protection, absorbing much of the cyclone's energy and safeguarding the geomorphic stability of the area.

Severe winds and waves during the cyclone destroyed the creeper cover, but the sand dunes underneath remained largely intact, protected from erosion. The vegetation absorbed much of the cyclone's energy, preserving the geomorphic stability. Sediment removed during the cyclone was trapped near the training walls at the Arasalar River mouth. Transects near the dunes and creepers showed less profile change post-cyclone, with recovery patterns emerging within a month. Sediment redeposition narrowed beach profiles, with Transect 1 reducing from 108.9 to 81.0 meters.

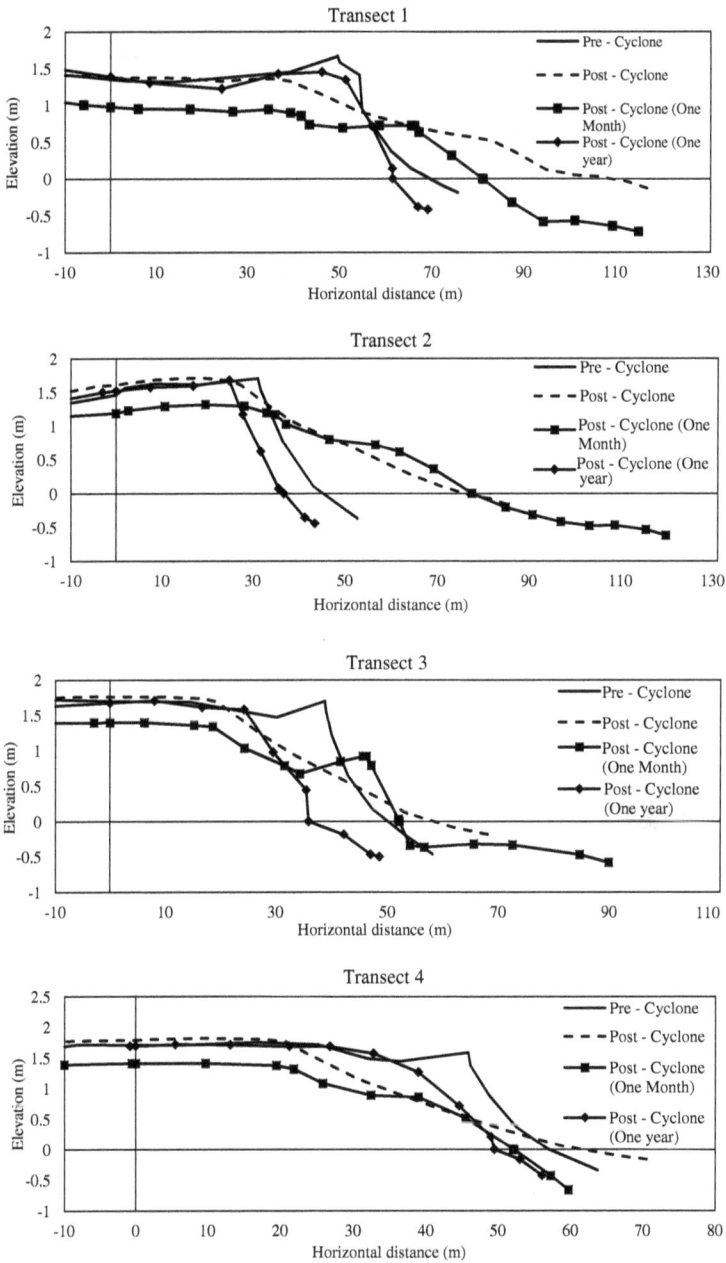

Fig. 11. Beach profile changes pre and post cyclone Gaja for all transects.

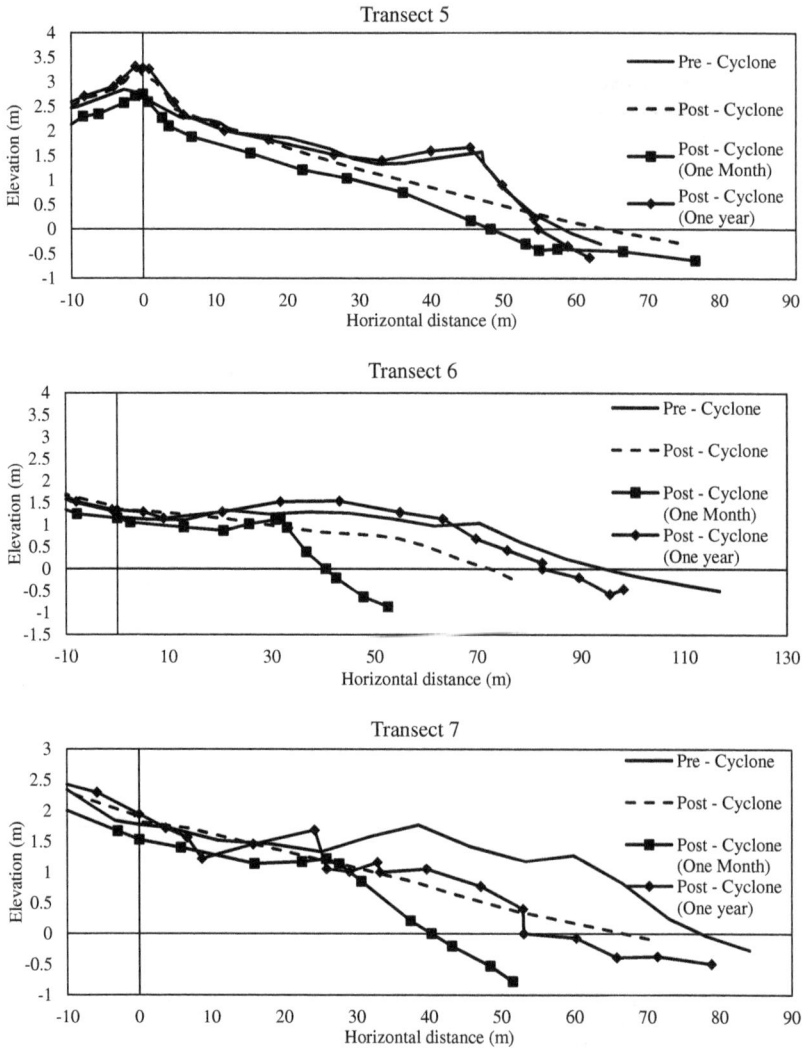

Fig. 11. (*Continued*)

Recovery was aided by natural sediment redistribution, protective dunes, vegetation, and sediment-retaining training walls.

By October 2019, one year after the cyclone, most transects showed significant stabilization and recovery. Transects 5, 6, and 7, near vegetation and stabilized dunes, not only recovered but exceeded pre-cyclone widths. For instance, Transect 6 expanded from about 40 m post-cyclone to about 83 m. This recovery was driven by vegetation stabilizing sediments and protecting the coastline. In contrast, more exposed transects showed less recovery.

In contrast, the more exposed Transects 1 and 2 exhibited slower recovery. While these profiles stabilized significantly within the year, they remained narrower compared to pre-cyclone conditions, with Transect 2 reducing from about 78 m one-month post-cyclone to about 37 m a year later. A combined effect of natural defenses and engineered structures is noticed in shaping the post-cyclone recovery process. The training walls not only prevented further erosion by trapping sediments but also provided localized stability to the beach profiles. Meanwhile, the vegetation cover and stabilized dunes acted as natural buffers (Figs. 12(b)), reducing wave energy, trapping sediments, and facilitating faster recovery. Though the floral cover was damaged by the strong winds and waves, its presence was instrumental in protecting the underlying sand dunes from erosion.

NIVAR cyclone

Thazhanguda (11° 46.002'N, 79° 47.575'E) is an estuary located in Tamil Nadu's Cuddalore district, a low-lying area frequently prone to floods and inundation. It is marked with a mean spring tidal range of 1 m (Mohapatra, 2015). Extreme weather events such as storm surges and associated large waves are generally encountered on the coast every year (Mohapatra, 2015). Soil types such as high sand and sandy loam play a crucial role in determining the impact of such events. This site is characterized by high

(a)   (b)

Fig. 12. (a) Beach cover pre-cyclone Gaja. (b) Beach cover post-cyclone Gaja.

sand and sandy loam with rich vegetation cover, including forests and coconut groves (Thirumurugan *et al.*, 2019). Four rivers — Ponnaiyar, Paravanar, Velar, and Cauvery — discharge into the sea at Cuddalore. The present study area, an estuary of the Thenponnaiyar River, has abundant vegetation cover and coconut plantations. A seawall built on both sides of the Ponnaiyar River mouth to prevent erosion has largely collapsed due to various extreme weather events. The river mouth is covered for most of the year, and the average annual rainfall in the Cuddalore district ranges from 105 cm to 140 cm (Thirumurugan *et al.*, 2019). The wave-dominated estuary is bounded by two rivers spanning a length of 2200 m. Sediment samples were collected as disturbed samples at all transects by removing the top layer of soil. These samples were then analyzed in the laboratory, following ASTM procedures for grain size analysis. The sediment sizes provided in the study represent average values, with the median grain size (D50) ranging from 0.3 to 0.5 mm.

The study area covers a shoreline length of 1000 m along the Thazhanguda village (11.7663874N and 79.7879067E) and 300 m in the vicinity of the river mouth. Considering the physical characteristics of the site, the beach profiles were collected at five transects. Transects 1 to 4 are located along the coast, and Transect 5 is at the river mouth. Transect 1 is in front of the vegetation cover, Transect 2 on open ground, Transect 3 within the coconut plantation, and Transect 4 on the south of the seawall. Transect 5 is at the north of the seawall. The figure depicts the transects' locations. The very severe cyclonic storm NIVAR made landfall along the east coast of India between Tamil Nadu and the Puducherry coast and it was about 50 km north of the present study area. This cyclone developed as a depression in the Bay of Bengal on November 23, 2020, intensified into a severe cyclonic storm, and crossed between Tamil Nadu and Puducherry (near 12.1°N and 79.9°E) between 2330 IST on November 25, 2020, and 0230 IST on November 26, 2020 (IMD, 2020). The post-cyclone field data (beach profile and bathymetry) were collected on November 28, 2020, while pre-cyclone data were collected on November 20, 2020.

Beach profiles were traced using RTK-GPS, integrating them with nearshore bathymetry to comprehensively quantify the coastal zone, including both the swash and surf zones. This combined profile helps understand the full impact on the coast, as changes occur both on land and underwater. RTK-GPS was also used to track the shoreline, enabling precise mapping from the river mouth to the habitat zone. The analysis focuses on changes in shoreline position and beach profiles due to cyclone landfall.

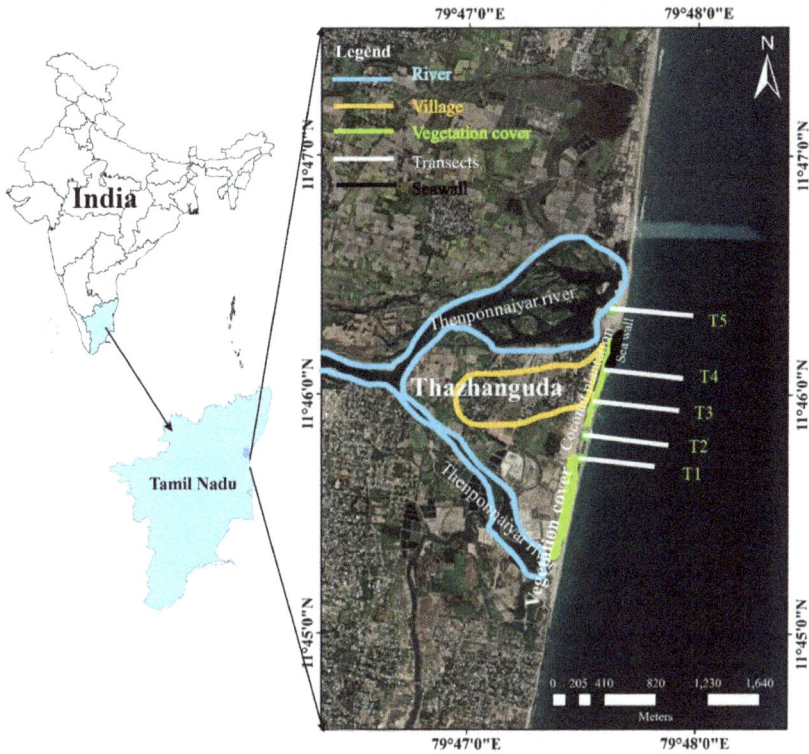

Fig. 13. Site description with the transects T1–T5 locations.

Shoreline changes

The changes in the shoreline due to the impact of cyclone in the vicinity have been projected onto a base map as depicted in 14. The representation shows that the shoreline underwent significant alterations. Before the cyclone, the shoreline from the habitat region (indicated as B) to the river mouth (indicated as A) was straight. However, the post-cyclone measurements showed that the area north of the seawall in the river mouth region eroded significantly. A recovery survey conducted in December 2020 revealed complete erosion at the river mouth Fig. 15(b) due to the continuous monsoon river flow that prevented siltation on the mouth. During this period, the coastal area south of the seawall (the habitat region, B) had further eroded than the post-storm scenario. The shoreline measurement three and twelve months after the cyclonic event was aimed at assessing the restoration capacity of the beach. Third-month survey revealed further erosion near the habitat region (Fig. 15) while the river mouth experienced accretion leading to its pre-cyclone scenario of mouth closure as mentioned earlier. This is due to the insignificant river flow during the fair-weather season and the less intense southerly littoral drift began. The accredited sediment was likely from the north due to littoral movement considering the closed nature of the estuary. However, rapid erosion was observed on the southern coastal stretch. In addition, scour was noted at several locations adjacent to the seawall. Overall, the river mouth returned to its pre-cyclone state within 75 days, while the southern stretch continued to experience erosion.

Seawalls significantly impact beach dynamics by altering hydrodynamic conditions and sediment transport, often aggravating erosion. While designed to prevent flooding and wave damage, they can cause "end effects," such as scouring at the base and "flanking" at the down-drift end, leading to additional erosion. For example, a seawall at Fansa, Gujarat (N20°20'24.684", E72°47'40.9704"), caused 20 meters of landward erosion within a year, with predictions of further erosion before stabilization (Balaji *et al.*, 2017).

Seawalls, while intended to protect the hinterland from erosion, often worsen existing issues. In the study area, aggressive erosion occurred near the seawall transect, leading to scouring at the toe stones and seawall damage. Additionally, beach loss resulted in damage to boat landing facilities along the coast.

Surge and wave heights are strongly influenced by the ocean bottom configuration and nearshore bathymetry. A slow-moving storm typically causes greater flooding in estuarine regions. **Cyclone Nivar** caused extensive flooding in Cuddalore, Kanchipuram, Chennai, and Pondicherry. On November 25, 2020, Cuddalore recorded 246 mm of rainfall and Pondicherry 237 mm, leading to inland flooding and increased river discharge, according to the Indian Meteorological Department (IMD).

The regions surrounding the Thenponnaiyar River, the second longest river in Tamil Nadu with a length of 500 km, were completely inundated, leading to significant crop damages and disrupted major road transportation. However, the inundation along the stretch with the seawall and vegetation was minimized. Coasts with gentle slopes and shallow water depths generally experience higher storm surges and smaller waves,

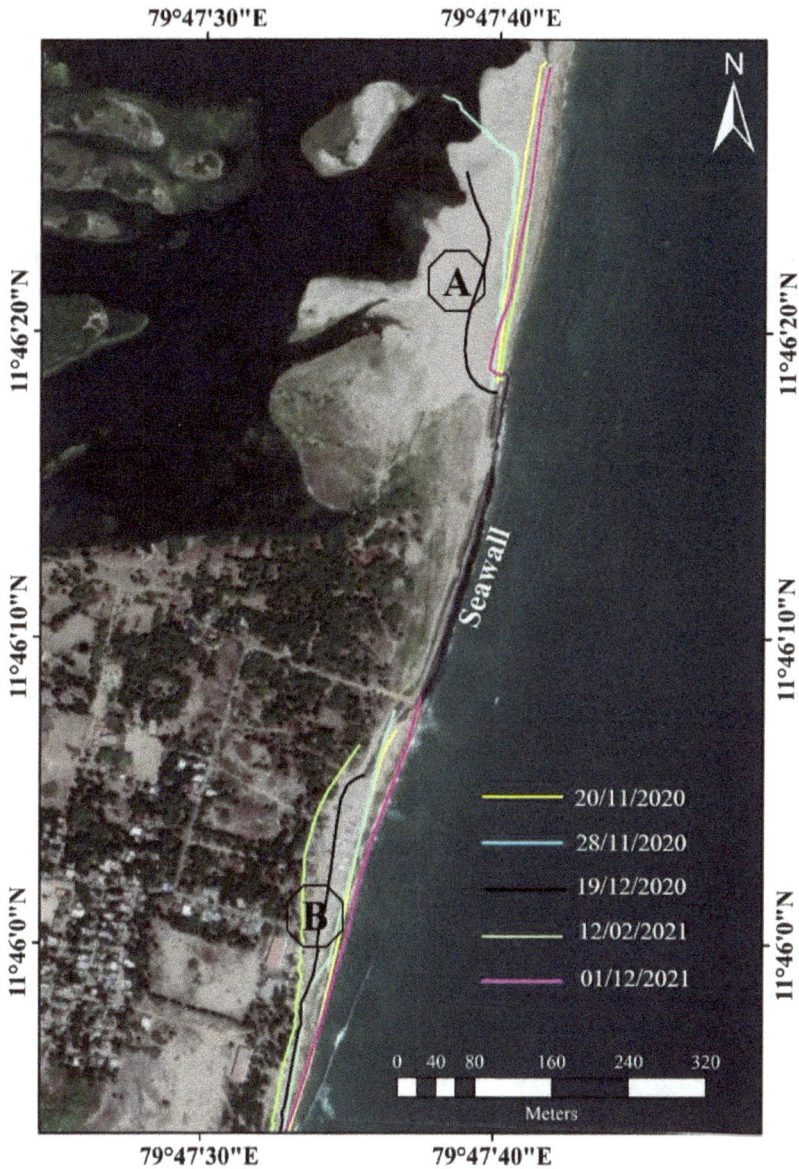

Fig. 14. Field-collected shoreline of Thazhanguda.

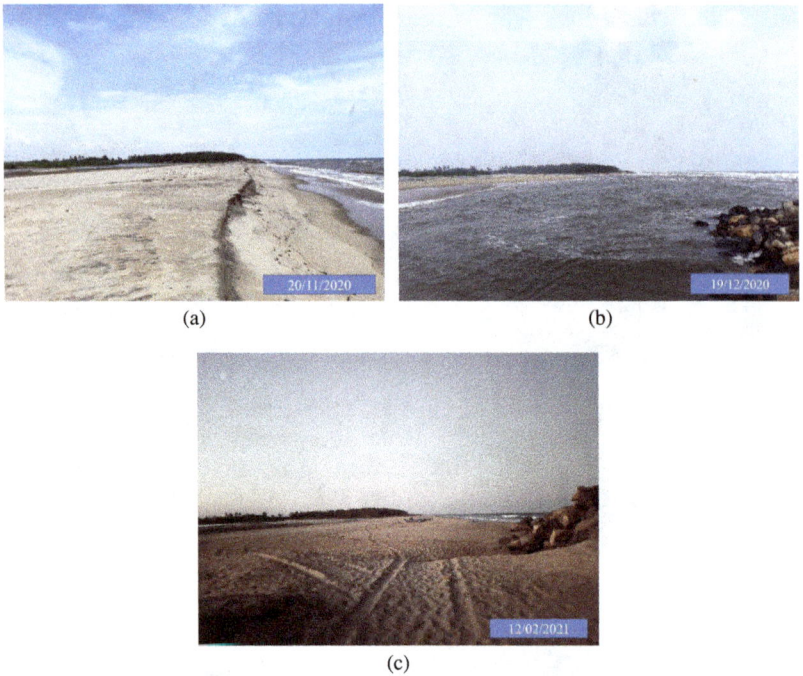

Fig. 15.   Impacts of Nivar cyclone on the river mouth of Thenponnaiyar River: (a) The pre-cyclone shoreline. (b) The shoreline along river mouth post-cyclone. (c) The siltation in the river mouth after three months.

Fig. 16.   The views of shoreline changes along the Thazhanguda village: (a) The eroded river mouth due to cyclone impact. (b) After 75 days of cyclone landfall.

whereas those with steep slopes are more likely to face lower storm surges and higher waves. The dense vegetation cover and abundant coconut plantations in the region have helped mitigate the large wave action. A study by Rao *et al.* (2013) on the VSCS Thane also observed that vegetation cover played a crucial role in protecting backshore areas from storm surges and run-ups. The run-up in this specific location was measured at +0.25 m, determined relative to watermarks on adjoining structures (Fritz *et al.*, 2007).

5.2.3 *Beach profile changes*

A distinct pattern of erosion and accretion was observed across Transects 1 to 4, highlighting the influence of the cyclone on the coastal morphology. Transects situated in front of vegetation cover experienced minimal changes, while Transect 2 in an open area underwent minor accretion. In a similar vein, Yin *et al.* (2019) have reported that areas near or covered with vegetation alleviated cyclone impacts more effectively. The role of vegetation in reducing inundation and wave height during the 2004 Indian Ocean tsunami was also discussed by Sundar *et al.* (2020) and Narayan *et al.* (2016) highlighting the benefits of nature-based defense systems in mitigating storm surge impacts. The measured beach profiles during pre- and post-cyclone are plotted in Fig. 17.

Transect 5, at the river mouth, exhibited drastic changes in terms of erosion and accretion, comparable to the other transects. The onshore part of the profile had eroded completely under the cyclone impact, with sediment loss ranging from an elevation of 2 m to a depth of 2 m, while the accretion was observed at depths from 2 m to 5 m. Given that this estuary is wave-dominated, sediment transport is influenced by wave intensity and direction. The estuary's morphology, crucial for local fishing activities, is dependent on the river mouth's dynamics. Such an opening of river mouths due to cyclonic impact is a common phenomenon, as also observed by Mohanty *et al.* (2020) on the Odisha coast during Cyclone Phailin.

Transect 1 exhibited a modest increase in sand volume with a net deposition of 60 m³/m. It may be indicative of a littoral drift deposit. Transect 2 showed a slight decrease in sediment volume with a subtle erosion of around 17 m³/m. It was relatively stable suggesting to have some

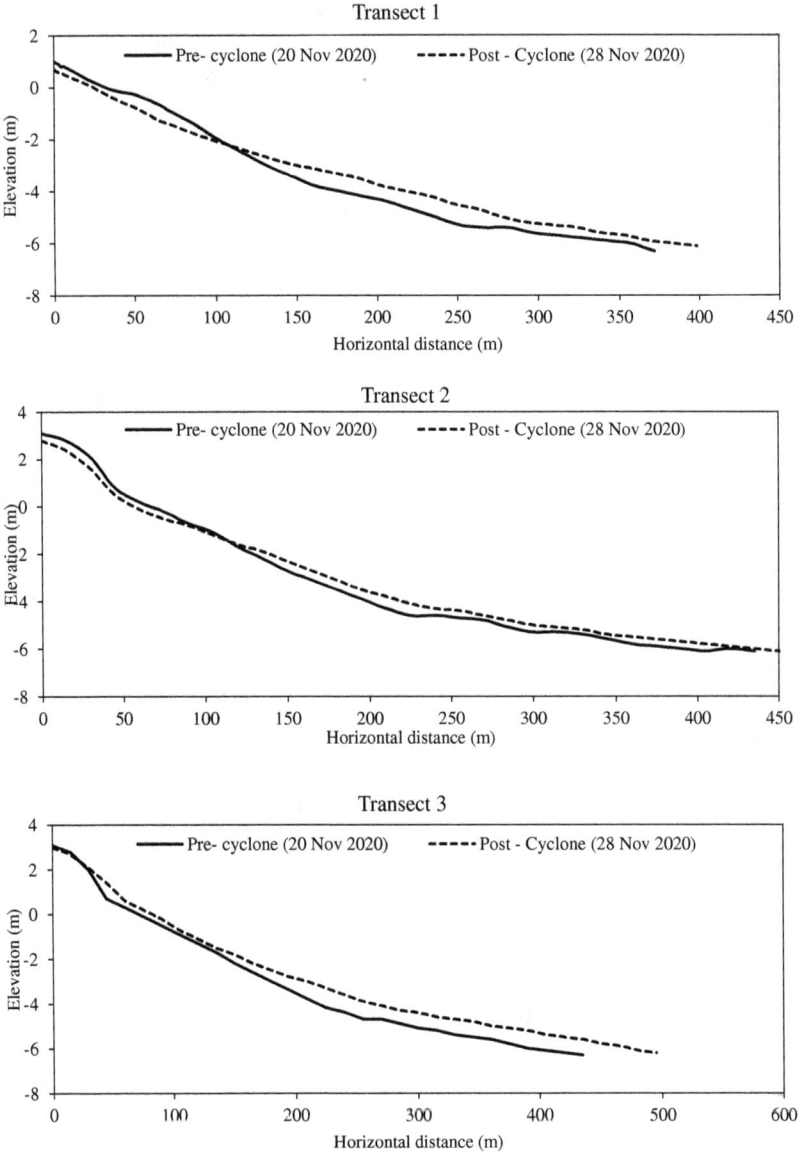

Fig. 17.   Beach profiles at five different transects along the coast.

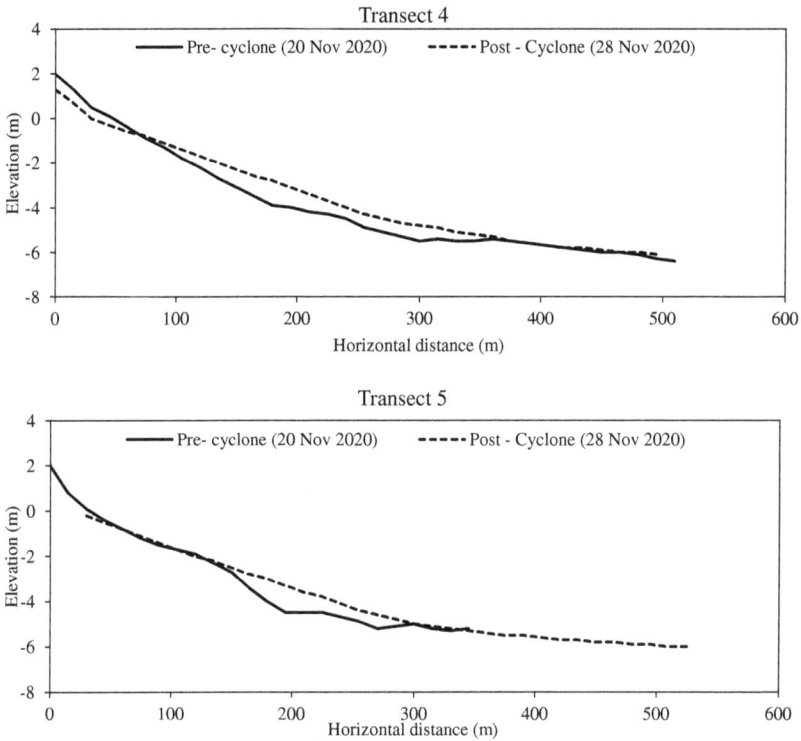

Fig. 17. (*Continued*)

resistance to more aggressive erosional forces of cyclones possibly due to natural land features. Transect 3 witnessed a significant increase in sand volume of 260 m³/m. This substantial accumulation points to a strong accretion, likely the redirected littoral drift during the cyclone, which gathered and deposited material in this location. Transect 4 experienced a notable reduction in sand volume of 97 m³/m. This indicates that the transect was subject to net erosion, where the cyclone's energetic waves and currents may have outpaced the littoral drift's capacity to replenish the lost sediment. Transect 5, positioned at the river mouth, showed the most considerable decrease, manifesting a loss of about 160 m³/m. This significant erosion aligns with that the cyclone's forces scoured the river mouth, possibly due to a combination of enhanced river discharge and storm surge, allowing the river to carry substantial amounts of sediment seaward. The analysis shows a complex sedimentary response in the Thazhanguda Estuary. Despite varying impacts across the transects, the overall volume increased, indicating a dynamic equilibrium within the estuary. This net gain in sediment, likely influenced by the cyclone's flooding and river discharge, features the estuarine system's capacity to adapt and retain sediments despite the severity of cyclonic events.

6. Coastal Zone Management

Coastal zone management (CZM) is critical for preserving coastal ecosystems, mitigating natural hazard risks, and ensuring sustainable development. One key strategy within CZM involves implementing policies that restrict or regulate construction in vulnerable zones, which are areas prone to natural hazards, such as flooding, storm surges, coastal erosion, or tsunamis, as well as those sensitive to environmental degradation, such as wetlands, dunes, and mangroves. Each country has its own CZM rules. Its implementation and strict adherence to them have been a challenge.

The most important constituents of Policies to Restrict Construction in Vulnerable Zones are as follows:

- Identifying Vulnerable Zones through risk assessment through GIS mapping and categorizing the different zones based on the magnitude and the frequency of the risks,
- Framework of Regulations zones which would include No construction or development within the coastal zone particularly in highly risk-prone zones,
- fixation of Setback Lines that defines a minimum distance from the shoreline within which construction is not allowed, considering erosion rates and projected sea level rise,
- zones that are permitted for construction for special activities like ports, power plants, etc. with stringent rules and regulations,
- Integration with Broader Coastal Management Plans including Climate Adaptation, i.e., align construction restrictions with climate adaptation strategies to address long-term changes such as sea-level rise,

- Disaster Preparedness through coordinating policies with disaster risk reduction plans to improve resilience.

By implementing and enforcing policies to restrict construction in vulnerable coastal zones, governments can create resilient and sustainable coastal communities, protect valuable ecosystems, and reduce the impacts of climate-induced and natural disasters.

7. Summary

Defining storm surge, its prediction and effect on the coast, and the significance of Integrated Approaches for mitigation have been discussed in this chapter. It is important to get prepared to face such challenges along the impact-prone coastal regions in the following perspectives:

Disaster Preparedness

- Early warning systems, powered by advanced predictive models, enable timely evacuations, significantly reducing casualties.

Coastal Resilience

- Accurate predictions inform the construction of durable coastal defenses and infrastructure to withstand extreme weather events.

Climate Adaptation

- Long-term planning based on predictive models helps communities adapt to future challenges posed by climate change.

Case studies always help in a better understanding of any subject matter. Hence, two case studies on the responses of the Karaikal and Thazhanguda coasts to two cyclones, viz., **Gaja** and **Nivar**, were considered to highlight the critical role of vegetation-based adaptations in mitigating the impacts of extreme weather events:

- Dense vegetation, such as creepers, coastal forests, and coconut plantations, acted as natural buffers, dissipating wave energy, reducing erosion, and protecting underlying sand dunes and backshore areas.

- Despite surface damage, vegetative covers ensured the stability of coastal features and facilitated rapid post-cyclone recovery through sediment trapping and stabilization.

The key takeaways of this chapter are as follows:

- Integrating vegetation-based strategies with engineered solutions enhances coastal resilience.
- Nature-based solutions not only mitigate immediate cyclone impacts but also support long-term recovery and ecological sustainability.
- With increasing cyclonic events due to climate change, preserving and restoring coastal vegetation is crucial for adaptive and sustainable management.

References

Balaji, R., Sathish Kumar, S., and Misra, A. (2017). Understanding the effects of seawall construction using a combination of analytical modelling and remote sensing techniques: A case study of Fansa, Gujarat, India. *The International Journal of Ocean and Climate Systems*, 8(3), 153–160.

Deltares. (2011). *Delft3D-FLOW User Manual*. Deltares, Delft, Netherlands. DHI (2021) (Danish Hydraulic Institute).

Fritz, H. M., Kongko, W., Moore, A., McAdoo, B., Goff, J., Harbitz, C., ... and Synolakis, C. (2007). Extreme runup from the 17 July 2006 Java tsunami. *Geophysical Research Letters*, 34(12).

Fukuoka, T., Katagiri, Y., and Kato, S. (2021). BIG-U Project's disaster reduction design implementation process and framework. *Journal of The Japanese Institute of Landscape Architecture*, 84(5), 587–590, March. DOI: 10.5632/jila.84.587.

Golda Percy, V. P., Sriram.V., Sundar. V., and Schüttrumpf, H. (2023). Effect of the buffer blocks in attenuating a tsunami-like flow. *Ocean Engineering Journal*, 286, 115489. https://doi.org/10.1016/j.oceaneng.2023.115489.

Jelesnianski, C. P., Chen, J., and Shaffer, W. A. (1992). SLOSH: Sea, lake, and overland surges from hurricanes. *NOAA Technical Report NWS 48*.

Van Alphen, J. (2015). The Delta Programme and updated flood risk management policies in the Netherlands. *Journal of Flood Risk Management*, May 2015. DOI: 10.1111/jfr3.12183.

Knabb, R. D., Rhome, J. R., and Brown, D. P. (2005). *Tropical Cyclone Report: Hurricane Katrina*. National Hurricane Center.

Knapp, K. R., Diamond, H. J., Kossin, J. P., Kruk, M. C., and Schreck, C. J. (2018). International best track archive for climate stewardship (IBTrACS) project, version 4. *NOAA National Centers for Environmental Information.* https://doi.org/10.25921/82ty-9e16.

Lal, M. and Harasawa, H. (2001). Future climate change and its impacts over small island states. Climatic Research Unit (CRU) Report. *International Journal of Climatology*, 21(14), 1765–1807.

Lokesha, Sundar, V., and Sannasiraj, S. A. (2013). Artificial reefs: A review. *International Journal of Ocean and Climate Systems*, IJOS, 4(2), June, 117–124. https://doi.org/10.1260/1759-3131.4.2.117.

Lokesha, Kerpen N. B., and Sannasiraj S. A., Sundar V. and Schlurmann T. (2015). Experimental investigations on wave transmission at submerged breakwater with smooth and stepped slopes. In *Proceedings of the Eighth International Conference on Asian and Pacific Coasts*, Chennai, India, September 7-10, 713–719. https://doi.org/10.1016/j.proeng.2015.08.356.

Lokesha, Sannasiraj, S. A., and Sundar, V. (2019). Hydrodynamic characteristics of a submerged trapezoidal artificial reef unit. *Proceedings of the Institution of Mechanical Engineers Part M: Journal of Engineering for the Maritime Environment*, 233(4), 1226–1239. https://doi.org/10.1177/14750902 18825178.

Luettich, R. A. and Westerink, J. J. (1992). ADCIRC: A parallel advanced circulation model for oceanic, coastal, and estuarine waters. *Technical Report*, Department of the Army, U.S. Army Corps of Engineers.

Mascarenhas, A. and Jayakumar, S. (2008). An environmental perspective of the post-tsunami scenario along the coast of Tamil Nadu, India: Role of sand dunes and forests. *Journal of Environmental Management*, 89(1), 24–34.

Mohanty, P. K., Kar, P. K., and Behera, B. (2020). Impact of very severe cyclonic storm Phailin on shoreline change along South Odisha Coast. *Natural Hazards*, 102, 633–644.

Mohapatra, M. (2015). Cyclone hazard proneness of districts of India. *Journal of Earth System Science*, 124, 515–526.

Mohammad, B. A. and Tanaka, H. (2016). Investigating the 2011 Tsunami Impact on the Teizan Canal and the Old River Mouth in Sendai Coast. Miyagi Prefecture; Japan. In *Tsunamis and Earthquakes in Coastal Environments*, April. DOI: 10.1007/978-3-319-28528-3_9.

Narayan, S., Beck, M. W., Reguero, B. G., Losada, I. J., Van Wesenbeeck, B., Pontee, N., ... and Burks-Copes, K. A. (2016). The effectiveness, costs, and coastal protection benefits of natural and nature-based defences. *PloS One*, 11(5), e0154735.

Perez, R. T. *et al.* (2016). Environmental impacts of Typhoon Haiyan in the Philippines. *Journal of Environmental Science and Management*, 19(1), 1–9.

Rao, V. R., Subramanian, B. R., Mohan, R., Kannan, R., Mageswaran, T., Arumugam, T., and Rajan, B. (2013). Storm surge vulnerability along Chennai–Cuddalore coast due to a severe cyclone THANE. *Natural Hazards*, 68, 453–465.

Regional Specialised Meteorological Centre Tropical Cyclones. Very Severe Cyclonic Storm "Gaja" over east-central Bay of Bengal (10–19 November): Summary. *India Meteorological Department*, New Delhi.

Sakthi Vasanth, N., Sriram, V., Sundar, V., and Schuttrumpf, H. (2024). A comparative study on the performance characteristics of buffer blocks configurations as energy dissipators. *Applied Ocean Research*, 153, 104202, ISSN 0141-1187. https://doi.org/10.1016/j.apor.2024.104202.

Sivakumar, K., Kumar, R. S., Ramesh, C., Adhavan, D., Hatkar, P., Bagaria, P., ... and Jyothi, P. (2016). *Conservation Strategy and Action Plan for the Marine Turtles and Their Habitats in Puducherry*. Wildlife Institute of India, Dehradun. 66.

Sundar, V. and Sannasiraj, S. A. (2019) *Coastal Engineering- Theory and Practice*. Advanced Series on Ocean Engineering, Volume 47, April 2019. https://www.worldscientific.com/worldscibooks/10.1142/11148, ISBN: 978-981-3275-91-1.

Sundar, V., Sannasiraj, S. A., Murali, K., and Sriram, V. (2020). *TSUNAMI: Engineering Perspective for Mitigations, Protection and Modelling*. Advanced Series on Ocean Engineering, June. ISBN: 978-981-121-606-0. https://doi.org/10.1142/11708.

Thirumurugan, P. and Krishnaveni, M. (2019). Flood hazard mapping using geospatial techniques and satellite images—A case study of the coastal district of Tamil Nadu. *Environmental Monitoring and Assessment*, 191, 1–17.

Very Severe Cyclonic Storm "NIVAR" over the Bay of Bengal (22nd–27th November 2020): A Preliminary Report, *IMD*, (2020).

Yin, K., Xu, S., Huang, W., Li, R., and Xiao, H. (2019). Modeling beach profile changes by typhoon impacts at Xiamen coast. *Natural Hazards*, 95, 783–804.

https://doi.org/10.1142/9789819816583_0006

Chapter 6

Characteristics of Tsunami Disasters in Japan

Yukiyoshi Hoshigami

*Department of Oceanic Architecture and Engineering.
College of Science and Technology, Nihon University, 7-24-1
Narashinodai, Funabashi-shi, Chiba, 274-8501, Japan*

hoshigami.yukiyoshi@nihon-u.ac.jp

About 50% of the population and 70% of the assets are concentrated in the coastal areas of the earth, and most of the places where culture and civilization flourished are "Port Towns" sited on ports. We discuss the current situation and issues of tsunami disaster prevention based on the actual situation of tsunami damage caused by the Great East Japan Earthquake in 2011, the Great Kanto Earthquake in 1923, and the Noto Peninsula Earthquake in 2024 from the viewpoint of urban development in coastal areas of Japan and hope that it will serve as a reference for safe and secure urban development in the future.

1. Introduction

In Japan, residential areas were formed by utilizing the experience of natural disasters until the early Edo period, but in the development of rice paddies from the 1600s onwards and the development of national land using modern civil engineering technology after the Meiji Restoration, convenience was pursued without considering the rules of thumb related to natural disasters.

In this paper, we focus on "Port Town" as an architectural space and from the viewpoint of tsunami disaster prevention town development. In addition, we focus on the history and disaster risk of the formation of Minato Town, the environmental changes as an architectural space as seen in the reconstruction status of the Great East Japan Earthquake, and the actual state of complex disasters, using the Great Kanto Earthquake of 1923 and the Noto Peninsula Earthquake in January 2024 as examples, and hope that these rules of thumb will help create a safe and secure coastal city.

2. What is a Tsunami Disaster?

There is a lot of knowledge about the mechanism of tsunami generation, and in this paper, we will omit the engineering mechanism of tsunami generation, but focus on the sudden changes in seafloor and coastal topography caused by earthquakes, volcanic activity, mountain collapse, meteorite fall, etc. As a mechanism of occurrence, it is believed that the rapid subsidence and uplift of the ground, such as the seafloor plate near the epicenter of the earthquake, causes the seawater on the seafloor to work together and cause water surface fluctuations.

The wavelength of a tsunami can span hundreds of kilometers, and the amount of seawater rushing to the coast is orders of magnitude larger than the waves. The propagation speed of the tsunami varies depending on the depth of the sea, and the speed is around 800 km/h (equivalent to that of an aircraft) offshore, but as the water depth becomes shallower, shallow water deformation occurs, the propagation speed decreases, and the tsunami height increases. Near the coastline, the speed is about 36 km/h (equivalent to the world record for the 100 m dash on land), the height is more than 10 m, and it is impossible to escape for human feet. Furthermore, the tsunami does not occur once, but several times after the earthquake, and the tsunami height is often higher after the second wave than the first wave.

For example, in plains, the water level gradually decreases inland, but if the plains are narrow and there are mountains or hills behind the coast, the tsunami will go upstream when it reaches the slope, and the seawater dammed on the slope will be increased. After reaching the maximum inundation depth, seawater tries to return from high to low in accordance with the laws of physics on Earth. This is called an "undertow" and drags drifting objects on land to the seaside at a high current speed. Images of the Great East Japan Earthquake also reported wooden houses being swept out to sea by undertows.

By the way, a natural disaster is a critical natural phenomenon (e.g., meteorology, volcanic eruptions, earthquakes, and landslides). The factors that cause natural phenomena to become disasters with human damage or the expansion of disasters are related to local social conditions. The vulnerability of the society (to disasters) is further increased by the lack of disaster prevention plans and the lack of appropriate crisis management, which increases the damage to people, the economy, and the environment.

For example, even if a natural phenomenon occurs, if there is no vulnerability in the area (e.g., no one lives in the area where the anomaly occurred), a natural disaster will not occur. The tsunami generated by the Tohoku Pacific Ocean Earthquake that occurred on March 11, 2011, was only a natural phenomenon, and the event that caused human casualties and economic losses as a result was the "Great East Japan Earthquake."

Based on the above understanding, the following chapter will focus not only on the Great East Japan Earthquake of 2011 but also on the Great Kanto Earthquake of 1923 and the Noto Peninsula Earthquake and Tsunami of 2024, and while focusing on urban development in coastal areas, we will delve deeper into the mechanism of disaster occurrence, deepen our understanding of disaster risks, and hope that it will help develop disaster prevention communities in the future.

3. Formation of Port Town in the Sanriku Coast

The Sanriku Coast, where the tsunami caused by the Tohoku earthquake struck, is a raised terrain with a total length of about 600 km from Samekaku, Hachinohe City, Aomori Prefecture to Manishiura, Ishinomaki City, Miyagi Prefecture. The water depth off the Sanriku coast is steep, and the Oyashio Current flows relatively close to land, and it is also known as the world's three largest fishing grounds. There are many port towns that have prospered from the fishing industry, such as Misawa, Hachinohe, Yamada, Otsuchi, Kamaishi, Ofunato, Kesennuma, Watana,

Onagawa, Ayukawa, Ishinomaki, and Shiogama from the north. In addition to sea urchins, oysters, scallops, and sea squirts, wakame seaweed farming accounts for 70% of the national production. On the other hand, about 300 km east of the coastline, the Japan Trench runs parallel to it. It is also a source of earthquakes and tsunamis (Ministry of Land, n.d.).

By the way, there are many shell mounds scattered in this area, and people who have sought their livelihood in the sea have lived in the sea since the Jomon period with the fishing gear excavated from the middens. Many of these middens have been excavated from high ground near the coast, and it is presumed that humans had no choice but to adapt to natural phenomena such as tsunamis as well as storm surges and high waves and that urban development was inevitably carried out according to the rule of thumb. For example, if you observe the fishing towns sited on the fishing ports on the Sanriku coast, you can see that the residential areas that have been located for a long time are public spaces, such as shrines and public halls, which are built exclusively on inconvenient and narrow land on high ground or steep slopes and which serve as places of mutual aid and public assistance. These are also located and concentrated in such high places.

In addition, the main house where the family lives is located on a hill, and the fishermen sleep in a boat shed near the beach during the busy season, which is a pioneer of the so-called separation of work and residence. Furthermore, in the survey after the East Japan Earthquake, most of the shrines, especially those with a history of more than 400 years since their construction, escaped damage, and the author's own research confirmed many cases of avoiding flooding at the last minute of the main shrine and the first torii gate and cases where local residents were saved by evacuating to the precincts.

On the other hand, in Japan, the paddy field policy was promoted during the Edo period, and efforts were made to improve the wetlands of the coastal area into rice paddies, and it flourished greatly as the kitchen of Edo (Ministry of Agriculture, n.d.), e.g., the plains of Rikuzentakata City in Iwate Prefecture and the cormorant residence in Kamaishi City. In addition, timber is easily available in Japan, wooden buildings have been adopted, and people have lived on well-drained hills and micro-highlands as a way to live in the hot and humid environment of the subtropics. However, river basins and coastal low-lying wetlands have high water levels, high humidity, and the risk of natural disasters such as storm surges and tsunamis, contributing to poor conditions for inhabited areas.

As a result, the population of Japan, which was 12.27 million around 1603 when the Edo shogunate was established, increased to 31.28 million around the time of the Kyoho reform (1716–1746), about 100 years later. In the 150 years up to the Meiji Restoration (1868), the number of people increased slightly to 33.3 million. Incidentally, in the 150 years since the Meiji Restoration, the number of people has quadrupled to 130 million (Long-Term Outlook Committee, 2011). After the Meiji era, the development of coastal areas accelerated further.

In other words, in the early 100 years of Edo, livelihoods centered on rice paddies developed, and as the population increased, people lived in low-lying areas away from Minato Town. However, in the Edo period, it was still a culture within walking distance, and industries and residential areas were concentrated along the line called the Old Hama Kaido. These Hama Kaido were located inland from the tsunami inundation area of the Great East Japan Earthquake, and paddy fields were an important land use for producing food and annual tribute. It is presumed that the paddy fields, which were wetlands, functioned as buffer zones against tsunamis, so it can be said that the rules of thumb and industrial structure of tsunami disasters that occurred periodically were eventually utilized in urban development until this period.

In addition, the Kitakami Mountains prospered from the time of the Fujiwara clan because of the abundance of gold veins. For example, Miyako Port was the outer port of the Southern Domain with Morioka as its residence and flourished as a military port and commercial port, and many Sengoku ships came and went offshore. Yamada, Otsuchi, Ofunato, Kesennuma, etc., were also natural ports that took advantage of the characteristics of the Rias topography and prospered as logistics bases. Currently, the Sanriku Coastal Road has been extended to the coast in addition to National Route 45, and in addition, the existence of the Sanriku Rias Line (Railway) has activated north–south movement, but the Sanriku Coast was called an isolated island on land until half a century ago. National Highway 45, which is the main trunk line, was opened in 1972. The entire line of the former Sanriku Rias Line (Railway) was opened in 1975. Until then, except for the east–west highway connecting the inland and Minato towns, the coastal movement was by boats and ships. In the 1700s, after the discovery of a large amount of magnetite at Sennin Pass, inland of Kamaishi, Kamaishi Port developed as an industrial port triggered by the construction of Japan's first Western-style blast furnace in the 1800s. Unosumai and Otsuchi Town, which are adjacent to

Kamaishi, developed as Kamaishi's best towns from the Meiji era to the period of high economic growth.

4. The Importance of Heuristics

Typical giant tsunamis that hit the Sanriku coast (Ministry of Land, n.d.) were the Sadakan earthquake tsunami (897), the Keicho earthquake tsunami (1611), the Meiji Sanriku tsunami (1869, 21.920 deaths), the Showa Sanriku tsunami (1933, 3,064 dead and missing), and the Chile earthquake and tsunami (1960, 142 people killed and missing). There are many books, press materials, and videos on the details of the tsunami damage, including the Great East Japan Earthquake, which are not specifically covered in this paper.

In Japan, until around the time of the Keicho earthquake and tsunami, no settlements were formed except for the area around Minato Town, except for the Hama Kaido and its surroundings. On the other hand, the Meiji Sanriku Tsunami, which had a death toll comparable to that of the Great East Japan Earthquake, was presumed to have damaged Minato Town, which flourished further by shipping during the Edo period, and coastal settlements built apart from the rule of thumb in the vicinity.

As for the aforementioned shrines, many of the shrines that were more than 400 years old escaped the disaster, while many of the villages with a relatively short history (developed after the Edo period) were damaged (Civil Engineering Research Center, n.d.). Many of these shrines were "separated" from the main shrine for daily worship. In other words, the damage situation of shrines built before the Keicho earthquake and tsunami (1611) and shrines enshrined in residential areas during the Edo period, when the population increased after that, suggests that the method of using the rule of thumb for tsunami risk in urban development changed after the Edo period.

Looking at the above from a bird's eye view, at least until the Edo period, Port Town was formed with fishing ports and harbors as the center of livelihood, and people with a strong sense of reverence for nature lived there. It is no exaggeration to say that in modern urban development, despite the fact that the Showa Sanriku Tsunami and the Chile Earthquake Tsunami and tsunami damage were subsequently experienced, the empirical rules were not fully utilized in urban development, and modern urban development in pursuit of convenience expanded, resulting in the current enormous damage.

In the field of engineering, there is an academic field called failure studies, and in Mr. Hatamura's book *Unprecedented and Unexpected* (Hatamura, 2011), there is a law in the decay of human memory, and after three years, it is forgotten at the individual level. In 60 years, it is forgotten at the local level, it disappears from the local community in 300 years, and after 1200 years, the culture does not know that the event occurred.

The rule of thumb that was handed down on the Sanriku coast was that it was a regional knowledge unique to the Sanriku coast, which strikes once every few decades, but the extent of the damage caused by the Chile earthquake and tsunami of 1960 was relatively small, and unfortunately, the rule of thumb has disappeared from the memory of Japanese organizations and regions that have achieved high growth for more than 60 years since the Showa Sanriku tsunami of 1933. As a result, it should be understood that this led to the enormous damage this time.

5. Tsunami Damage Caused by the Great Kanto Earthquake

Here, as an example of modern tsunami damage, we will take the Great Kanto Earthquake that occurred about 100 years ago on September 1, 1923, and comprehend the difference from the tsunami damage caused by the Great East Japan Earthquake.

5.1. *Overview of the Great Kanto Earthquake*

Of the total 105,000 dead and missing, 92,000 (87%) were from fires and 11,000 (11%) were from house collapse, accounting for 98% of the total number of victims. In addition, most of the socio-economic losses, such as 1.9 million people affected, more than 210,000 houses burned down, and 160,000 partially destroyed houses, were caused in urban areas, making it the first urban catastrophe that Japan had experienced.

However, at the time of the disaster, former Prime Minister Tomosaburo Kato had died, and Gonbei Yamamoto was in the cabinet, and the leaders of the government did not realize the seriousness of the situation until the night of September 1, when the earthquake occurred, and the response began in earnest on the morning of Monday, September 3, after the cabinet was formed. In addition, since the earthquake occurred on a Saturday when the employee was working a half-day shift, many of the employees went home immediately after the earthquake and did not report for work.

The military began relief operations immediately after the disaster at the discretion of each unit. Immediately after the earthquake, the news and communication agencies ceased to function, but at that time, wired telegraphs and telephones were the main communication tools, and radio communication was not a stable source of information, and the Army Air Corps conducted communication and reconnaissance activities. The main means of communicating information to the citizens were printed materials and messengers distributed to various places. Radio broadcasting did not start until two years after the earthquake, indicating the difficulty of collecting and sharing information at that time.

In this way, the Great Kanto Earthquake was a catastrophe that exceeded the expectations of people at the time, and disaster preparedness was neglected due to overconfidence in technological progress, and the damage expanded.

At present, Japan has experienced many large-scale disasters such as the Great Kanto Earthquake, the Great Hanshin-Awaji Earthquake, and the Great East Japan Earthquake, and has prepared materials and systems for emergency response, including earthquake resistance and fire prevention of buildings.

5.2. *Overview of tsunami damage*

In the Great Kanto Earthquake, buildings collapsed and fires caused enormous damage, and a lot of the information found was about relief and relief efforts in urban areas (Cabinet Office, n.d.). For example, Yokohama, which is close to the epicenter, was an international port at the time and had many foreigners staying there, so it received a lot of foreign support. However, there is little recorded information in areas other than urban areas, such as the Sagami-nada coast and the Boso Peninsula. In addition, because it is difficult to separate the damage caused by tsunamis and earthquake ground motions, the reports on tsunamis are fragmentary, and the overall picture and the actual damage are not clear.

In addition to the description that the tsunami struck from Sagami Bay to the Izu Peninsula and caused major damage, including more than 1,000 casualties, there is also a description that the death toll from the tsunami was 200–300 and the death toll from landslides was 700–800 and that most of the deaths occurred in Kanagawa Prefecture. The actual number is unknown, but this does not include deaths due to the collapse of houses. As for the tsunami height, the maximum wave height was about 1 m on the Tokyo coast and there was no damage from the tsunami, but the main tsunami height on the Sagami Bay coast was 6 m in Atami City, Shizuoka Prefecture, Miura in Kanagawa Prefecture, and 9 m locally in Yuigahama in Kamakura City, and more than 300 people were missing, a tsunami of 5–7 m reached the coasts of Zushi, Kamakura, and Fujisawa, and the tsunami reached the Yuigahama stop of the Enoshima Electric Railway (near the current Hase No. 4 level crossing).

Also, it was 9.3 m in Aihama (present-day Tateyama City) in Chiba Prefecture and 8 m in Susaki. The Boso Peninsula has observed uplifts of 1.5 m or more, and the truth of these figures is unknown. It is characterized by the fact that it was 12 m in Atami and Izu Oshima (Okada) and more than 9 m in Ito and Tateyama (Suzaki). The tsunami height of 12 m in Atami was the result of surveying based on the testimony that it reached less than 2 m from the top of the Hanazuki pine (a pine tree of about 10 m high at 4 m above sea level). The tsunami height indicates the water level near the coastline, and it cannot be ruled out that the tsunami height at that time may have included the height of the tsunami at that time.

By the way, in the same report (see Ref. 7), the weather before and after the earthquake is described as follows: First of all, rainfall occurred from the western part of Kanagawa Prefecture to the eastern part of Kanagawa Prefecture from August 31, the day before the earthquake, to the morning of the earthquake, and landslides occurred frequently mainly in areas with a seismic intensity of less than 6 or higher immediately after the earthquake. According to the Great Kanto Earthquake Photographs Collection (Japan United News Agency, 1923), the mountains along the coast of Hayakawa collapsed and buried hundreds of towns and villages of rice paddies. The Hakone Electric Railway was completely buried by a landslide, the entire line of the Atami Railway (line up to Manazuru at that time) collapsed, most of the line (railroad track) collapsed into the sea due to landslides, and the entire section was disrupted.

Based on the aforementioned information that 7–8% of the deaths in Kanagawa Prefecture were caused by landslides, it was reported that landslides caused more severe damage. It is understandable that the tsunami was not taken up as a topic. In addition, the sea conditions were bad on the day of the earthquake. There is a description that 100 beachgoers were missing in the tsunami at Yuigahama Beach in Kamakura City, but on September 1, the day of the earthquake, the weather was bad and the waves were high, suggesting that there were few people bathing in the sea, so the actual situation is unknown. In addition, there was damage to the Boso Peninsula (Chiba Prefecture General Affairs Department, 2009). At that time, the investigators dispatched by the government only recorded a rough survey conducted by a few people, and the details of the damage are unknown.

It is noteworthy that the tsunami arrived quickly. In Atami and Nefu River on the Izu Peninsula, a tsunami with a height of 5–6 m swept in 5 minutes after the main shock. In the vicinity of Kamakura, the tide receded by 2–300 m immediately after the earthquake, and in about 10 minutes, waves about 2–3 times the size of the earthen wave swept in. The tsunami was repeated 2–3 times, and the second wave was larger. The first wave reached the coast of Tateyama in 10 minutes.

6. History of Tsunami Defense Along the Sanriku Coast

In this section, the history of the construction of tsunami protection facilities along the Sanriku coast since modern times is summarized. In Japan, seawalls were built near the coastline as a countermeasure against storm surges to protect the hinterland. Especially on the Sanriku coast, the frequency of tsunami attacks is high, so seawalls were built as tsunami counter measures even before the Great East Japan Earthquake. However, only the area around Hironomachi on the northern coast of the Sanriku coast had seawalls for tsunami protection behind the fishing ports and harbors in this area, and the southern side of the Sanriku coast, especially Minatomachi in the south of Miyagi Prefecture, at Taneichi Fishing Port in Hirono Town, a tsunami seawall of about T.P. +12 m was built between the harbor and fishing port and the hinterland based on the assumption of damage from the Showa Sanriku tsunami (Fig. 1).

In general, ports and fishing ports have breakwaters to ensure the tranquility of the port, so damage from waves and storm surges is less likely to occur. Moreover, the idea of tsunami protection behind the port was hardly dealt with before the Great East Japan Earthquake, and in Minatomachi, which was built and developed after the modern era, there was no plan or concept to protect the residential area behind the port. Iwate Prefecture, which has the Sanriku coast, has experienced the Meiji Sanriku, Showa Sanriku, and Chile earthquakes and tsunamis since the beginning of modern times and has been promoting the development of

Fig. 1. Tsunami seawall at Taneichi Fishing Port.

tsunami seawalls as a storm surge countermeasure project. Especially in the coastal areas where there was no flood damage from the Showa Sanriku and Chile earthquake and tsunami, the main Minato town was developed after World War II, so no protective lines were set up in the hinterland of the port, and the tsunami seawall was not built.

On the other hand, according to a survey conducted by the Ministry of Land, Infrastructure, Transport and Tourism on the state of seawall maintenance after the Great East Japan Earthquake, the newly constructed section of the seawall was built, e.g., in almost all ports in Iwate Prefecture, such as Miyako, Kamaishi, and Ofunato, a new tide barrier line (planned location of tsunami seawall corresponding to level 1 tsunami) was newly installed between the port and the hinterland after the Great East Japan Earthquake. This is because it was believed that it could be defended by a breakwater at the mouth of the bay.

In other words, even if the damage estimate changed due to the Great East Japan Earthquake and hindered convenience, it can be said that the priority has changed to protect Minato Town.

In addition, in relatively large ports, the damping effect of the breakwater against the external force of the tsunami is exerted, so the top height of the seawall corresponding to the level 1 tsunami, based on the inundation simulation, is lower than that of the outside port. Tsunami seawalls of this height will also be built behind the harbor and fishing port.

On the other hand, in Kamaishi Bay, where the breakwater at the mouth of the bay is located, both the port and the coastline have a planned top height of T.P. +6.10 m, but in Karatan Bay, which is adjacent to the south side, a tsunami seawall with a planned top height T.P. +14.50 m will also be built behind the port. In Kamaishi, the height of buildings in Minato Town, such as on the Sanriku coast, is standard at about two stories, and if you actually go to both areas, you will not feel much discomfort at Kamaishi Port, but if you go to Miyako Port, you will feel a slight discomfort due to the presence of a wall that protrudes from the height of the standard buildings in the town.

In addition, since the port and the residential area have been separated by a new protective line, it is necessary to thoroughly examine and discuss the existing convenience, impact on livelihoods, cost-effectiveness, and socio-economic perspectives.

7. Examples of Reconstruction after the Great East Japan Earthquake

In this section, we introduce the current state of reconstruction centered on the protective lines and present the following observations as an engineer, while focusing on the livelihoods of ports and fishing ports and the residential areas behind them.

7.1. *Case 1: Example of a tsunami seawall (Rikuzentakata City Fishing Port)*

Rikuzentakata City, Iwate Prefecture, is a fishing port located on the western shore of Hirota Bay. Small-scale fishing port facilities centered on several outboard motor vessels are lined up in each cove on the Rias coast, and behind each fishing port facility is a village that clings to a steep slope. Before the Great East Japan Earthquake, there was an existing seawall (upright seawall) with a T.P. +4.5 m behind the fishing port, but after the earthquake, four tsunami breakwaters (upright embankments) with a height of T.P. +12.5 m were constructed at the position blocking the entrance to the Rias terrain behind the fishing port (Fig. 2).

If you go up to the top of the seawall and look at the hinterland, as shown in Fig. 3, there are still houses that were not flooded by the Great East Japan Earthquake, but there are no reconstructed houses in places

Fig. 2. Tsunami seawall in the affected area.

Fig. 3. Residential area seen from the top of the tsunami breakwater.

Fig. 4.   Landside of the tsunami seawall.

below the top height, i.e., there are no residents in the protection area of
the tsunami seawall, and vacant lots are spreading.

In addition, as shown in Fig. 4, the tsunami seawall has a wall height
of about 7–8 m, and although it has an opening due to a land lock, there
is a feeling of pressure when standing on the land side of the facility. In
addition, the air was felt to be clear due to the stagnation of air compared
to the surrounding area. In terms of the architectural environment of the
hinterland settlement, not only the change in the view (obstruction) but
also physical factors such as wind and humidity may change, and in some
cases, the life of the wooden building may be affected. In addition, as
pointed out in many previous studies, it cannot be ruled out that the sense
of security provided by the hardware facilities may contribute to the disaster risk in the event of a tsunami evacuation.

7.2. Case 2: Relocation to higher ground (Kamaishi City)

This is an example of the relocation of fishing villages to higher ground
in the towns of Unosumai and Otsuchi in Kamaishi City, Otsuchi Bay,
Iwate Prefecture (Figs. 5 and 6). On the other hand, in the Nebama area

Fig. 5. Aerial view of the left bank of the Uju River and the hinterland.

Fig. 6. Hinterland seen from the top of the seawall.

on the right bank side, the T.P. +20 m or more, which is also suitable for the level 2 class tsunami, was relocated to higher ground at the request of the residents, and the existing coastal seawall (storm surge response, T.P. +5.60 m) and fishing port facilities were restored (Figs. 7 and 8). Standing on the local high ground, the inconvenience of living on a hill away from

Fig. 7.   Aerial view of the Nehama area on higher ground.

Fig. 8.   Residential area on high ground in the Nehama area.

the fishing port is inevitable, but the view opens up, and above all, it is a residential area with a sense of security against tsunamis.

In the town of Otsuchi, which is located on the northern shore of Otsuchi Bay, a tsunami seawall with a height of T.P. +14.50 m was built behind the fishing port to protect the urban area. In the Akahama area, which is located at the eastern end of Otsuchi Town, the area was

Fig. 9. Seawall and relocation to higher ground in the Akahama area.

Fig. 10. Seawall and hinterland in the Akahama area.

relocated to higher ground, as in the Nehama area, and the residential area was secured against an L2 tsunami. However, as shown in Figs. 9 and 10, a level 1 class-compatible tsunami seawall with a top height much lower than the residential height (T.P. +6.40 m) is built behind the fishing port, but there is no protection target behind this tsunami seawall, and the intention of the protection plan cannot be found.

Fig. 11.   Aerial view of the tsunami seawall in the Sukuura area.

7.3. *Case 3: Example of different measures in adjacent bays (Karakuwa Town, Kesennuma City)*

The Maine and Sukuura districts of Karakuwa Town, Kesennuma City, Miyagi Prefecture, are small bays located at the base of the Kesennuma Bay side of the Karakuwa Peninsula and are good natural harbors that are difficult for waves from the open sea to penetrate (Fig. 11).

On the other hand, in the Sukuura area on the east side, an L1 seawall (T.P. +9.90 m sloping embankment) was built about 100 m inland from the fishing port (Fig. 12). It escaped flood damage and sits as it was before.

7.4. *Example of integrated maintenance of tsunami seawall and building (Minami Town, Kesennuma City)*

Located in the innermost part of Kesennuma Bay in Miyagi Prefecture, Minamimachi is a port town that formed and developed the central urban area as a logistics base for Kesennuma and has a ferry terminal that connects remote islands. Although the urban area near the port was flooded by the Great East Japan Earthquake, since the area made a living from the port, many discussions and proposals were made over time, and various ideas were made to separate the port and the town with a tsunami seawall in the reconstruction plan.

Fig. 12. Aerial view of the relocation to higher ground in the Maine area.

Fig. 13. Floating flap gate installed at the top of the seawall at Kesennuma Port.

For example, by using a movable flap gate at the top of the seawall (Fig. 13) and raising the height of the ground behind the seawall, it was devised so as not to impair the view of the seaside from residential areas in the hinterland, pedestrians, and passing vehicles. By placing the buildings of the terminal facilities in close proximity to this upright embankment and integrating it with the tsunami seawall, the protective function is devised so that it does not interfere with livelihoods (Figs. 13–15).

Fig. 14.   Embankment vegetation for access to the seaside of the seawall and landscape considerations.

Fig. 15.   Buildings integrated with seawalls.

Fig. 16. Buildings integrated with seawalls.

The first floor of the terminal facility is a pilot's-style parking lot, and convenient facilities are provided on the second floor and above, and cafes and restaurants are located higher than the top of the seawall (Fig. 16).

7.5. Case 5: Example of loss of boundaries due to tsunami seawall (*Ogatsu Town and Minamisanriku Town*)

Ogatsu Town in Miyagi Prefecture is a prosperous place in Minato Town, but after the Great East Japan Earthquake, In this way, efforts to relocate the town itself to the land side were actually practiced over time in several areas such as Fudai Village in Iwate Prefecture and Yoshihama in Kamaishi City after the Meiji Sanriku Tsunami, and it is worth mentioning that there was almost no substantial damage in these settlements during the Great East Japan Earthquake.

On the other hand, Minatomachi is a place of residence centered on livelihood, and a residential area where those who do not make a living cannot survive. Currently, if you go to Ogatsu Town, a tsunami seawall with a top height of T.P. +9.70 m has been developed around the seaside of the Rias Coast, but there is no sign of people in the lowlands behind the seawall. Settlements have disappeared, and such areas are scattered along the Sanriku coast.

As pointed out in the reconstruction review of the Great Hanshin-Awaji Earthquake, it is important to discuss in advance a reconstruction plan that takes into account long-term urban development because disaster reconstruction cannot be applied to anything other than the mechanism and technology at the time of the disaster.

8. A Qualitative Study of Urban Development after the Great East Japan Earthquake

Since the Great East Japan Earthquake, the term "giant seawall" has come to be used frequently, but when people say "huge," they probably mean something that we cannot handle, something that is scary (too big).

Although it is qualitative, based on the results of field surveys such as the cases described in the previous section, we will organize miscellaneous impressions about the changes in the architectural space and the built environment after the earthquake reconstruction from the standpoint of an engineer.

8.1. Changes in the Architectural Residential Environment

Long-term monitoring, including actual measurements, is necessary for wind stagnation and increased humidity due to the construction of tsunami seawalls, deterioration of living comfort, including psychological aspects, inconvenience of access, and impact on livelihoods.

8.2. Promotion of disaster prevention risks

In order to reduce the risk of disaster prevention by constructing a tsunami seawall, it is necessary to reconsider local knowledge, including empirical rules, and the risk of tsunami disaster prevention in the next few decades, around hundreds of years (grandchildren and children's generations), should also be evaluated.

8.3. *Sustainability of local communities*

In the future, it is necessary to scrutinize whether it has become possible to run a sustainable local community, including the livelihood and the economic and industrial indicators of Minato Town.

8.4. *Increased maintenance and management costs*

Buildings can be dismantled when they finish their role and reach the end of their life. Tsunami seawalls must maintain and manage their functions on the order of several hundred years to come. In particular, facilities such as reinforced concrete, equipment, and machinery deteriorate as the years pass, so there are concerns about the increase in costs for maintenance and renewal.

8.5. *Safety, security, and cost-effectiveness*

While wasteful and inefficient facility arrangements can be seen, there are also many areas that have realized ideal urban development in terms of safety, such as relocation to higher ground.

9. Complex Disasters Caused by the Noto Peninsula Earthquake

At 4:10 p.m. on January 1, 2024, the "Noto Peninsula Earthquake" occurred with an epicenter on an active fault off the western coast of the Noto Peninsula with a magnitude of 7.6. In addition to the collapse of many wooden houses and tsunami damage, there was also a lot of damage due to ground uplift and liquefaction.

According to the Ministry of Land, Infrastructure, Transport and Tourism website (Ministry of Land, 2004), 245 people were killed, 8,528 houses were completely destroyed, and 19,408 houses were partially destroyed. Around 145 sections of national and prefectural roads were closed due to landslides and slope collapses, and about 190 hectares of tsunami flooding occurred in the three municipalities of Suzu City, Noto Town, and Shiga Town, and the tsunami reached in as little as 2 minutes, reaching a maximum inundation depth of about 4 m. In addition, more than 10,000 cases of residential land damage occurred due to liquefaction in a wide area from Ishikawa Prefecture to Niigata Prefecture.

The author conducted a damage survey over two days from April 20 to April 21, 2024. Here, we report on the current situation, focusing on the damage to buildings four months after the earthquake. In addition, there was almost no damage around Toyama Station and Kanazawa Station, and the lives of citizens vicinity were normal.

9.1. *Summary of survey results*

(1) **April 20 (Sat) 7:10, Ota, Takaoka City, near Amaharu Beach (seismic intensity 5 or higher), Fig. 17:** A part of the cliff in the back mountain had collapsed, but there was no damage to the building or road.

(2) **April 20 (Sat) 10:38, Noto-cho Matsunami Fishing Port (seismic intensity 6 strong), Fig. 18:** The ice-making facility at the fishing port was tilted, but the damage to other buildings was minor. However, cracks and cave-ins were seen on the quays, water strikes, roads, etc., of the fishing port, and simple repairs were made.

Fig. 17.   Ota, Takaoka City, near the Amaharu Coast.

Fig. 18. Matsunami Fishing Port, Noto Town.

(3) **April 20 (Sat) 10:54, near Koiji Beach, Koiji, Noto Town (seismic intensity 6 weak), Fig. 19:** Damage to interlocking scattering, volcanic sand eruptions, and damage to coastal facilities were confirmed due to the collapse and liquefaction of the exterior walls of the hotel.

(4) **April 20 (Sat) 11:27, Ukai, Suzu City, near Mitsuke Island (seismic intensity 6 or higher), Fig. 20:** In the vicinity of Mitsuke Island, a famous tourist spot, most of the houses in the residential area collapsed. Relatively new houses were also destroyed. Debris was concentrated on both sides of the road, vehicles could barely pass, and signs were already given that the risk level had been assessed. On the other hand, houses with functional bracing were spared from collapse. In some places, manhole hatches were raised by more than 1 m due to liquefaction. Around the seaside park on Mitsuke Island, traces of tsunami inundation with a height of about 1 m were seen, seawall facilities were destroyed due to undertows, and sea erosion cliff collapses were confirmed on Mitsuke Island due to earthquake motion.

Fig. 19. Noto-cho Koiji near Koiji Coast.

Fig. 20. Near Mitsuke Island, Suzu City.

Fig. 21. Near Kasugano-otsu, Takaritsu Town, Suzu City.

Portable toilets and mobile homes were brought in for the accommodation of reconstruction support workers.

(5) **April 20 (Sat) 11:54, near Kasugano-Oto, Takaratsu-cho, Suzu City (seismic intensity 6 or higher), Fig. 21:** Damage to buildings was seen in almost the entire area. The first floor of the house that had escaped collapse was destroyed by the tsunami. There were no traces of the fishing gear warehouse near the coast, leaving only the concrete foundation, but there were traces of the house being swept inland by the tsunami after the house collapsed, and traces of tsunami overflow and drifting were also seen in the river.

(6) **April 20 (Sat) 12:20, Sugawara Shrine (seismic intensity 6 or higher), Fig. 22:** At the site of the collapse of the shrine hall of Sugawara Shrine, it was a traditional wooden building made of beautiful black shiny "Noto tiles (roof construction)," but it could not withstand the weight of the tiles and completely collapsed.

(7) **April 20 (Sat) 15:28, Kawai-cho, Wajima City, near the morning market (seismic intensity 6 strong), Figs. 23 and 24:** The center of Wajima City, which recorded a maximum seismic intensity of 6 or

Fig. 22.   Sugawara Shrine, Nonoe Town, Suzu City.

Fig. 23.   Kawai-cho, Wajima City, near the morning market.

Fig. 24. Kawai-cho, Wajima City, near the morning market.

higher, was severely damaged, and the closer it was to the center. The road was very undulating and difficult to drive. Traces of collapse and liquefaction of old wooden houses, fire sites in the morning market in wooden constructed areas, and the collapse of seven-story buildings caught my eye in the scenes I saw on the news every day. This seven-story building was built in 1972, but the ground near the foundation was locally uplifted, with the bottom of the footing visible, and all the support piles were missing. The details are unknown because the national government and academic surveys have not been conducted, but the site situation and several RC 3–7-story buildings over 50 years old exist in the city. It was reported that it was tilted by up to 4 degrees, suggesting the possibility of damage due to liquefaction and lateral flow.

April 21 (Sun) 10:57, Near Osaki, Kahoku City (seismic intensity 5 or higher), Fig. 25: This is an area where lagoons (salt marshes and lagoons) formed on the land side of coastal sand dunes were reclaimed and

Fig. 25.   Near Osaki, Kahoku City.

reclaimed, and the Ministry of Land, Infrastructure, Transport and Tourism Land Preparation "Liquefaction Ease Map. (Ministry of Land, n.d.)" However, the risk of liquefaction was 2–4 (represented by the risk level 0 (lowest) to 4 (highest)), and the entire town suffered liquefaction damage extensively and continuously, such as telegraph poles tilting heavily, roads waving, liquefaction and lateral flow occurring in low-altitude residential areas on the land side of sand dunes, houses tilting heavily, and steps of up to several meters between roads and residential lots.

9.2. *Characteristics of the damage*

(1) **Damage caused by earthquake ground motion:** Regarding the earthquake resistance of wooden houses, buildings constructed after the enforcement of the revised Building Standards Law in June 2000 had load-bearing walls such as those bracing firmly bonded to the foundations and beams during the earthquake and were spared from

collapse even in the Kumamoto earthquake with a seismic intensity of 7. According to the knowledge of the general public, "wooden houses are fine if they are 'new and earthquake-resistant' as amended in 1981." In addition, since December 2020, there have been multiple earthquakes with a seismic intensity of less than 6 in the same area, and the most recent earthquake on May 5, 2023 also damaged the wooden house. It is desirable to increase the number of options for protecting lives, such as the introduction of shelters in living rooms and bedrooms, as well as making the entire house earthquake-resistant.

(2) **Liquefaction damage:** After World War II, especially during the period of high economic growth, in recent years, hazard maps have been released, and the danger of such places has been foreseeable in advance. Although there are a few cases where liquefaction damage itself leads to human casualties, it is clear that the disruption of lifelines, including buried objects, takes a considerable amount of time to recover, as evidenced by the cases of liquefaction-affected areas in the Great Hanshin-Awaji Earthquake and the Great East Japan Earthquake. It is desirable to discuss a preliminary reconstruction plan, including relocation from the viewpoint of urban development. In addition, the collapse and slope of RC buildings in Wajima City once again demonstrated the vulnerability of the foundations of old RC buildings.

(3) **Ground uplift and landslides:** In this earthquake, a ground uplift of 1–4 m was observed in the coastal area. The uplift of the ground led to the reduction of tsunami flood damage described later. On the other hand, the fishing village area connected by the coastal road was isolated by land due to many landslides, leading to delays in rescue and transportation of supplies. In addition, the uplift caused the fishing port to dry up and the quay to leave the water, making it impossible to rescue and provide support from the sea route. The same problem was pointed out in the Great East Japan Earthquake, but many of the fishing ports on the Noto Peninsula are small-scale fishing ports. In addition, the fisherman population has been declining and aging for some time, and it will be necessary to have a thorough discussion on how to rebuild the village as a fishing village, including cost-effectiveness.

(4) **Tsunami damage:** Before the disaster, the tsunami hazard map of the Noto Peninsula (Ishikawa Prefecture, n.d.) predicted a minimum tsunami arrival time of 2 minutes. The tsunami actually arrived within

2 minutes, and the expected inundation range of the hazard map and the inundation results after the field survey were generally in agreement, and it was as predicted in advance. Although no areas were flooded for a long time, traces of inundation, such as tsunami drift, were confirmed near the collapsed houses, and the houses collapsed after the earthquake.

According to the announcement by Ishikawa Prefecture, the death toll from the tsunami is 2 (the number of people with the consent of the bereaved families), but there have been several reports of deaths with evacuation routes blocked by collapsed houses underneath. The maximum inundation depth of the tsunami is about 4 m, and there are cases where vertical evacuation to the second floor could have been saved if the building had not collapsed due to earthquake motion, but in the situations where it collapsed or was underneath, it was not possible to evacuate, and they drowned.

In the Great Kanto Earthquake about 100 years ago, about 1,000 people were confirmed to have died in the tsunami, but many of the areas saw houses collapsing before the tsunami hit, and the number of tsunami deaths has not been accurately ascertained.

9.3. *Observations*

This was a short survey, but it was especially important to see the collapse of houses and liquefaction. All of the events on the Noto Peninsula could have been foreseen in advance, and there is a very high possibility that the coastal areas where the Tokai and Tonankai earthquakes and earthquakes directly under the Tokyo metropolitan area are at risk of being hit by similar complex disasters in the future.

10. Preparing for Future Tsunami Disasters

In the future, the declining birthrate, aging population, and depopulation will surely improve in Japan, and it is estimated that the population will be 37.70–64.07 million in the 2100s, 100 years after the next giant tsunami is expected.

In order to create a city that balances disaster prevention and livelihood in coastal areas, it is important to realize safe, secure, efficient, and effective planning by utilizing the long-standing rule of thumb, "local knowledge," in addition to planning that takes into account population decline.

In general, many people think of tsunami evacuation as "to higher ground" or "far away (outside the flooded area)," but in principle, "vertical evacuation" is the rule. "To higher ground" is the right choice if there is a higher ground nearby, but not necessarily if it is in a remote place. The aforementioned evacuation actions, based on the experience of disasters along the coast of the Izu Peninsula and the Pacific coast of Tateyama City, Chiba Prefecture, are based on high ground behind them.

References

Cabinet Office. (n.d.). Report of the Expert Committee on the Inheritance of Lessons Learned from the Disaster: The Great Kanto Earthquake of 1923 [2nd report]. http://www.pa.thr.mlit.go.jp/kamaishi/nigiwai/sanriku-history/index.html (Retrieved July 28, 2023).

Chiba Prefecture General Affairs Department, Fire and Earthquake Disaster Prevention Division. (2009, March). Disaster prevention magazine: The Great Kanto Earthquake.

Civil Engineering Research Center, Nagisa Research Laboratory. (n.d.). Nagisa Research Laboratory. http://www.pwrc.or.jp/nagisa12K.html (Retrieved June 1, 2021).

Hatamura, Y. (2011). *Unprecedented and Unexpected: Learning from the Great East Japan Earthquake*. Kodansha.

Ishikawa Prefecture. (n.d.). Ishikawa Prefecture tsunami inundation area map. https://www.pref.ishikawa.jp/bousai/tsunami/index.html (Retrieved May 12, 2024).

Japan United News Agency. (1923). Great Kanto Earthquake charts. Taisho 12. National Diet Library Digital Collections. https://dl.ndl.go.jp/.

Long-Term Outlook Committee, Policy Subcommittee, National Land Council. (2011, July 21). Interim summary of the long-term outlook for the national land. Ministry of Land, Infrastructure, Transport and Tourism.

Ministry of Agriculture, Forestry and Fisheries. (n.d.). Miyagi: Irrigation heritage sites. https://www.maff.go.jp/j/nousin/sekkei/museum/m_izin/miyagi/ (Retrieved June 1, 2021).

Ministry of Land, Infrastructure, Transport and Tourism. (2024). Disaster and disaster prevention information: Damage and response to the Noto Peninsula Earthquake in Reiwa 6. https://www.mlit.go.jp/saigai/saigai_240101.html#n0 (Retrieved May 12, 2024).

Ministry of Land, Infrastructure, Transport and Tourism, Tohoku Regional Development Bureau, Kamaishi Port Office. (n.d.). Sanriku history. http://www.pa.thr.mlit.go.jp/kamaishi/nigiwai/sanriku-history/index.html (Retrieved June 1, 2021).

Ministry of Land, Infrastructure, Transport and Tourism, Tohoku Regional Development Bureau. (n.d.). Tohoku Regional Development Bureau. https://www.thr.mlit.go.jp/ (Retrieved June 1, 2021).

Ministry of Land, Infrastructure, Transport and Tourism, Regional Development Bureau. (n.d.). Map of ease of liquefaction in Ishikawa Prefecture. https://www.hrr.mlit.go.jp/ekijoka/ishikawa/ishikawa.html (Retrieved May 12, 2024).

www.ingramcontent.com/pod-product-compliance
Lightning Source LLC
Chambersburg PA
CBHW050543190326
41458CB00007B/1895